Well Placement Fundamentals

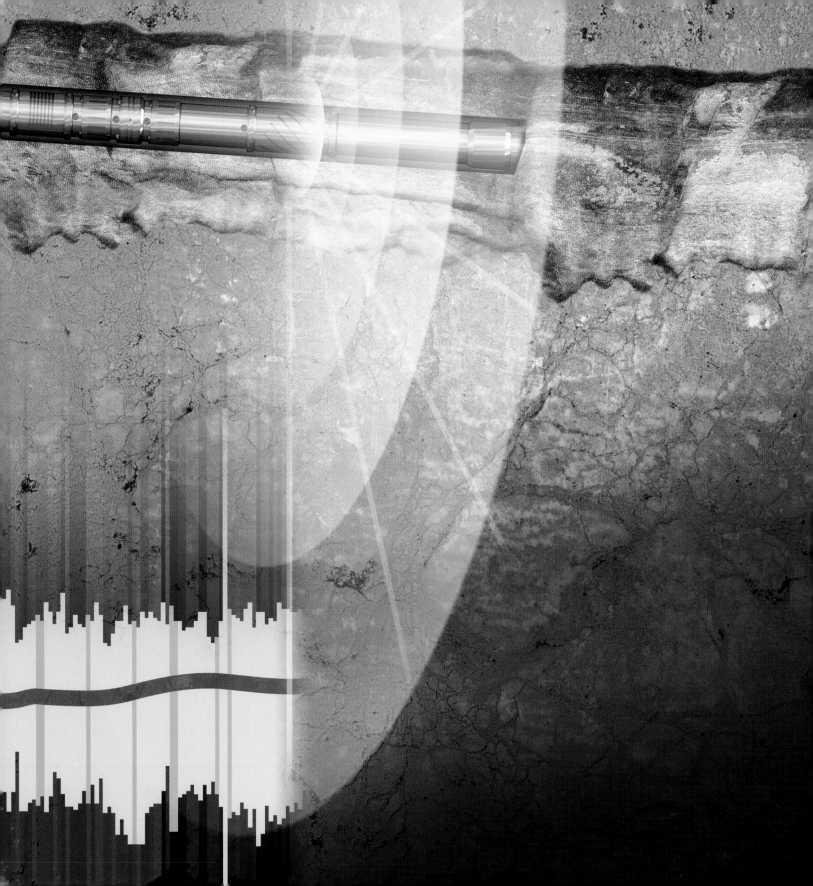

Well Placement Fundamentals

Roger Griffiths

Schlumberger

Schlumberger

225 Schlumberger Drive

Sugar Land, Texas 77478

www.slb.com

ISBN 978-097885304-4

09-DR-0050

Contents

Well placement is the planned interactive positioning of a wellbore using geological criteria and real-time measurements.

1.1 Definition

Well placement is the planned interactive positioning of a wellbore using geological criteria and real-time measurements. The well placement process is an interactive approach to well construction, combining technology and people to optimally place wellbores in a given geological setting to maximize production or injection performance. Accurate well placement helps improve the return on the money invested in drilling the well.

1.2 Introduction

The success of a well can be measured in both the short and long term:

- In the short term the success of the well is determined by whether it is drilled safely, efficiently, on time, and on budget and is producing hydrocarbons at the expected rate or better.

- In the long term considerations such as access to reserves, delayed onset of water production, extended production, and reduced intervention costs determine the total revenue generated from the well and hence the return on investment from drilling the well.

Well placement improves both the long-term and short-term performance of a well (Fig. 1-1). The drilling rate of penetration (ROP) is generally improved because the well remains in the more porous reservoir, which can be drilled faster than the surrounding formation, and sidetracks are avoided. By staying in the reservoir rather than the nonproductive surrounding formations, production is also improved.

The key to maximizing reserves recovery is placement of the well in the reservoir such that it produces hydrocarbons for the longest possible time and drains the formation as completely as possible. By accessing pockets of untapped formation and avoiding unwanted fluids, a well ultimately delivers the maximum return on investment by delivering the maximum possible hydrocarbon volume with the minimum associated water or unwanted gas production.

An example is the placement of a well at the top of a reservoir with an active aquifer, which drives the hydrocarbon to the top of the formation as production proceeds (Fig. 1-2). Placing the well at the top of the reservoir (blue well path) enables hydrocarbon production to continue for a long period before the onset of any associated water. Attic oil, left behind between a horizontal wellbore (black dashed well path) and the top of the reservoir, is minimized because the well is placed as close to the cap of the reservoir as possible. Both short-term production and long-term reserves access are maximized.

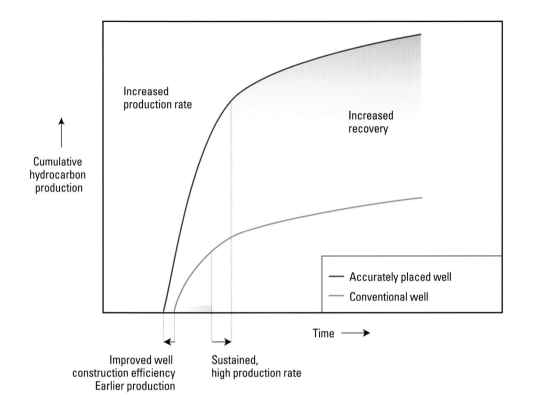

Figure 1-1. Well placement
improves well and asset
economics by improving well
construction efficiency, delivering
higher sustained hydrocarbon
production rates, and improving
hydrocarbon recovery.

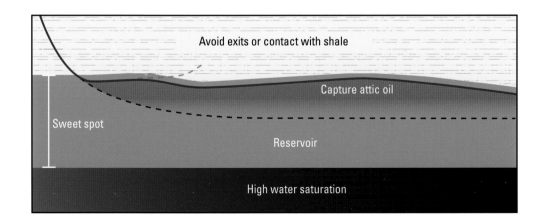

Figure 1-2. Optimal well
placement (blue solid line)
ensures short-term production,
long-term reserves access,
and delayed onset of water
production. The red dashed
line shows an undesirable,
unintended exit into the overlying
shale. Drilling too low in the
formation (black dashed line)
is also undesirable because
it leaves attic oil when the
formation is drained.

1.3 Well placement economics

Horizontal wells are more expensive to drill than vertical wells. Yet the horizontal proportion of the total number of wells drilled each year continues to increase. This is because, when drilled in the correct place, horizontal wells enhance wellbore productivity in both the long and short term, as outlined in the previous section.

Horizontal wells generally have higher risk than vertical wells. The requirement to keep the well in the productive interval demands improved directional drilling and reservoir understanding beyond that required to drill vertically into a target. Where these improvements are not in place prior to drilling horizontally the increased expenditure may not deliver improved well performance because the target zone is not exposed by the well as assumed in the well prognosis.

A number of operating companies have experimented with horizontal drilling only to be disappointed with the subsequent well performance. Typically these initial wells have been drilled geometrically, without the benefit of well placement techniques to ensure that the well is placed in the target zone. In many cases initial disappointment has resulted in reluctance to continue with horizontal drilling, despite the significant improvements in productivity and hence profitability that are possible.

The well placement techniques outlined in the following chapters help mitigate the risks associated with positioning the well in the target. By improving drilling efficiency, reservoir contact, productivity, and profitability, well placement can deliver the horizontal well potential that has eluded some operators.

Horizontal wells remain more expensive than vertical wells. Adding well placement services to the authorization for expenditure (AFE) further increases the initial well cost. But simple engineering fundamentals imply that the majority of reservoirs will yield higher production and recovery with horizontal wells than with a similar number of vertical wells. Cost-benefit analyses generally favor horizontal drilling, provided the risks associated with wellbore positioning in the target are mitigated through the application of appropriate well placement techniques.

Accurate well placement helps improve well construction efficiency, reduce drilling risk, extend reservoir contact, maximize reservoir exposure, improve well performance, and enhance ultimate hydrocarbon recovery. Figure 1-3 shows how these factors can improve reservoir development economics and help maximize the return on investment from drilling a well.

1.4 Well location terminology

The placement of a well requires that its position in 3D space be clearly defined. Three coordinates are required to describe the location of any point in a wellbore in 3D space. The most common in oilfield practice are the following:

- true vertical depth (TVD)—the vertical depth of the wellbore independent of its path (Fig. 1-4). In the case of a vertical well, measured depth (MD) is the same as TVD.

- displacement—the shortest distance from the surface location of a well to the projection of the bottom of the well (or other point in the well) to the Earth's surface (Fig. 1-5).

- azimuth—the angle between the displacement line and true north or magnetic north measured in a horizontal plane, typically measured clockwise from north.

Figure 1-3. Return on investment in drilling a well is dependent on both the rate and duration of production.

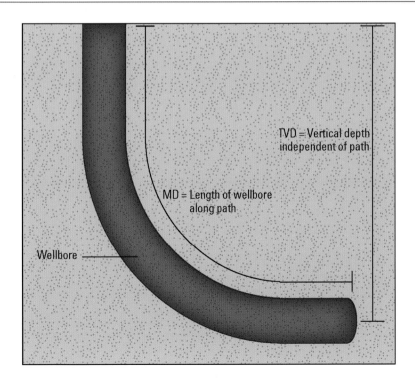

Figure 1-4. True vertical depth is the vertical depth of the well independent of the well path.

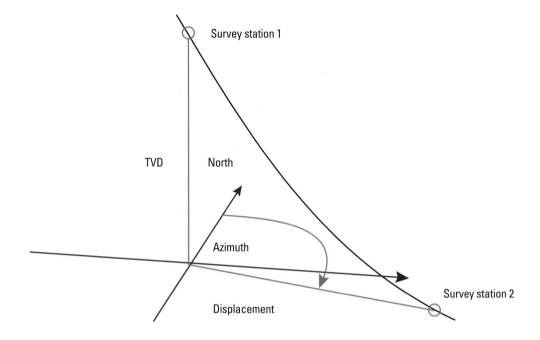

Figure 1-5. The position of a point in space, such as survey station 2, can be defined by its TVD, displacement, and azimuth. Displacement is the shortest horizontal distance between two points in the well. It is generally referenced to the surface well location.

Well position information is generally presented in a 2D format using plan views (looking down from above) and vertical sections (looking horizontally at a projection of the wellbore in a vertical plane). Figure 1-6 shows a typical pair of 2D representations of a well trajectory. The planned well (black line) and the well as drilled (blue line) are shown in side view perpendicular to the planned well and in plan view from above. In this example the drilled well has departed from the plan in both TVD and azimuth. Plots such as this are updated regularly while drilling to enable the driller to compare how close the drilled well is to the well plan.

The plane of the vertical section is generally along the planned well azimuth, but it can be created at any azimuth. This is demonstrated in Fig. 1-7, where different vertical section lengths are shown depending on whether the vertical section plane is taken along the planned well azimuth (green vertical section length) or an alternative azimuth (brown vertical section length). Care must be taken when looking at vertical section views to ensure that the azimuth of the section is known because the vertical section length of a well varies depending on the azimuth to which it is projected. In an extreme case, a horizontal well drilled perpendicular to the section view azimuth appears as a point, giving no indication of the length of the well.

The true horizontal length (THL) of a well is the length along a projection of a trajectory to the horizontal plane. THL is independent of well and vertical section azimuths because it follows the changes in well azimuth.

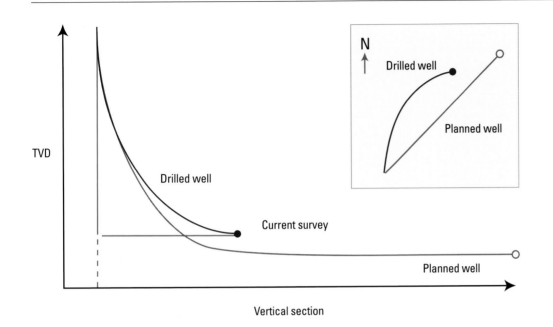

Figure 1-6. Well trajectory information is generally presented in vertical and horizontal projections called the vertical section view and plan view, respectively.

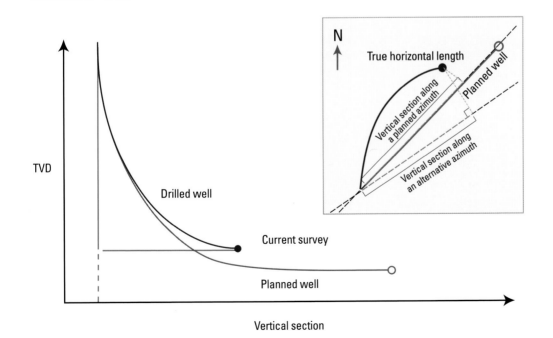

Figure 1-7. The vertical section lengths shown in plan view differ depending on the azimuth on which the drilled well trajectory is projected. The true horizontal length of the wellbore is independent of the well and vertical section azimuths.

With current technology it is not possible to measure TVD, displacement, or THL directly, so they are computed from survey data acquired using the direction and inclination (D&I) sensors in measurement-while-drilling (MWD) tools. For further information about how these measurements are made, refer to Chapter 4, "Measurement-While-Drilling Fundamentals."

Three measurements are used to define a survey:

- measured depth—the along-hole length of the trajectory, essentially the length of drillpipe fed into the wellbore

- inclination—the deviation from vertical, irrespective of compass direction, expressed in degrees that is measured initially with a pendulum mechanism and confirmed with MWD accelerometers or gyroscopes

- azimuth—the angle between the displacement line and a north reference (generally magnetic, true, or grid north) in a horizontal plane, typically measured clockwise from north.

A well drilled to its final depth is said to have reached total depth (TD).

1.5 The three complementary methods of well placement

There are three complementary methods of well placement:

- Model, compare, and update is the original method, which involves modeling log responses based on a formation model and well trajectory, comparing the modeled responses with real-time measured logs, and updating the formation model to match the real-time measured logs. This method can be applied to any real-time log data.

- Real-time dip determination requires formation data from opposite sides of the wellbore, preferably images scanned from the entire inner circumference of the wellbore, that are transmitted while drilling. Formation dip is calculated by the correlation of features from opposite sides of the borehole. The dip is then extrapolated away from the borehole and the well steered on the assumption that the formation dip does not change significantly.

- Remote detection of boundaries for real-time well placement currently requires the deep azimuthal electromagnetic measurements of the PeriScope* distance-to-boundary service. Through an inversion process, the distance and direction to changes in formation resistivity can be determined. Well placement using this technique requires knowledge of the resistivity boundaries within a reservoir sequence of layers and which of those boundaries the measurements and inversion can detect.

Traditional well placement involves specification of a geological target or targets for which drilling engineers then design a well trajectory to intersect and follow. In engineering a planned trajectory, drilling engineers must consider factors such as avoidance of nearby wells (anticollision); hole cleaning, removal of cuttings, and formation pressure control (hydraulics); the ability to manipulate the drillstring without exceeding drillpipe torque and tensile limits (torque and drag); and the ability to steer the well (bottomhole assembly [BHA] tendencies). Flexibility must be built into the design of a well to allow for the possibility that the targets could require modification during drilling. Target locations are generally selected based on a reservoir structural model derived from seismic data and well-to-well log correlation. Owing to the limited vertical resolution of surface seismic data and the assumptions made during the well-to-well log correlations, the actual subsurface structure is often different from that indicated by the model. The upper panel of Fig. 1-8 shows an example where three targets have been selected and a well planned to intersect all the targets. However, the formation was not as smooth and continuous as expected but had subseismic faulting that, if drilled according to the original geometric plan, would have resulted in the well having little exposure to the reservoir. Through the monitoring of real-time data and application of well placement principles, the actual well (red line in the lower panel) achieved greater reservoir exposure and hence production than planned.

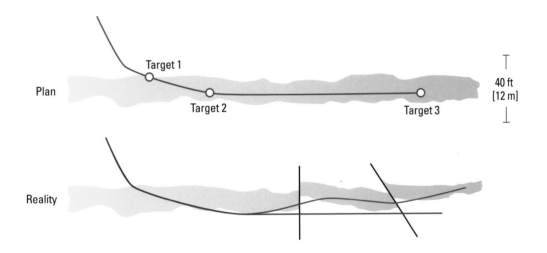

Figure 1-8. Traditional well placement involves drilling geometrically through targets specified by geologists, but the geology is often different from expectations.

It is important to remember that all data has associated uncertainty. The limited vertical resolution of surface seismic data and the assumptions made during well-to-well log correlations result in formation models with uncertainty in the TVD of formation tops and lateral changes in both formation dip and thicknesses. Even subsurface measurements such as the position of the wellbore computed from surveys have associated uncertainty. Each survey station has an ellipsoid of uncertainty associated with the uncertainty in the MD of the survey and the accuracy of the magnetometers and inclinometers used for the survey measurements themselves. Further information on surveying uncertainty can be found in Chapter 4, "Measurement-While-Drilling Fundamentals." Because the position of a well is computed from these surveys, the accumulation of uncertainties from each of the survey stations results in an expanding cone of uncertainty along the path the well is expected to follow with a given probability (Fig. 1-9).

Figure 1-9. The cone of
uncertainty of the well location
results from the accumulation
of uncertainties from each of
the survey stations used in
calculation of the well location.

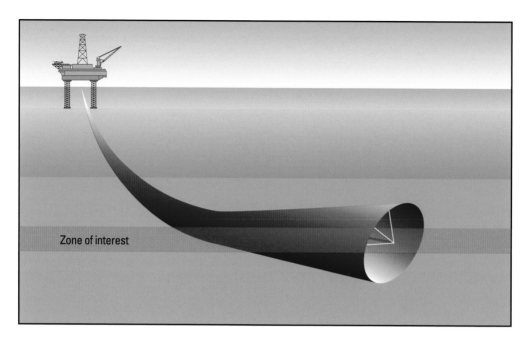

Zone of interest

1.5.1 Model, compare, and update

The model, compare, and update technique requires that a structural model of the formation of interest is built
and populated with the formation parameters that will be measured in real time. These formation parameters
are typically gamma ray (GR), resistivity, density, and neutron response. The real-time measurements are
made with logging-while-drilling (LWD) tools and the data transmitted to the surface in real time using an
MWD system. The MWD surveys are used to define the position of the well trajectory in 3D space and the
corresponding point on the formation model. With the LWD tool position and the properties of the surrounding
layers known, the theoretical response of the LWD measurement can be computed and compared with the
measured response from the downhole tool. If they match, then the model gives a reasonable representation
of the borehole position relative to the surrounding layers. If they differ, then the formation model must be
adjusted to give a match between the theoretical LWD response computed from the model and the measured
response from the downhole LWD tools.

1.5.1.1 Building the formation model

The formation model incorporates both the structural information and the properties of each layer. The first
step in constructing such a model is to acquire the properties over the interval of interest from a representative
offset well. Figure 1-10 shows how an offset well log is squared and propagated into a geological model
to create a property model. Note that the value of the formation property is represented by the color of the

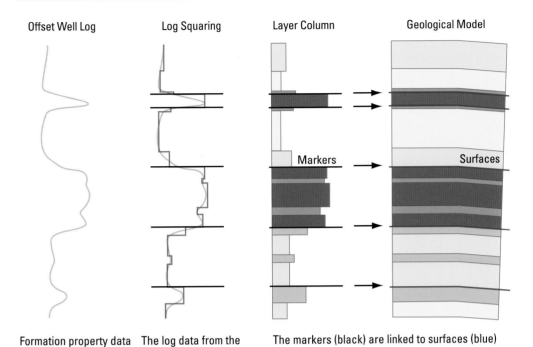

Offset Well Log   Log Squaring   Layer Column   Geological Model

Markers   Surfaces

Formation property data (GR, density, neutron, resistivity) from a representative offset well is obtained.

The log data from the offset well (green) is squared (red) to define the layer boundaries and formation markers (black) are identified

The markers (black) are linked to surfaces (blue) in the geological model. The squared log values are propagated across the curtain section between the surfaces.

Figure 1-10. Construction of the property model involves squaring the offset well logs and propagating the formation properties into a geological structure model.

layer. In Fig. 1-10 higher formation property values are represented by darker colors. This approach allows representing the formation property value represented in a 2D formation cross section, as shown in the right panel of the figure.

Building the formation model starts with using the deflections in a log from the offset well to define the layer boundaries. The density log is recommended for this task because it has the best vertical resolution of the commonly available triple-combo logs. The value of each formation parameter is then defined for each layer in a process known as log squaring. The resulting layer column is a squared representation of a formation property (GR, resistivity, density, neutron, photoelectric factor [PEF]). To check that the squared log is a good representation of the offset well log, a forward model of the appropriate logging tool used in the offset well is applied to the squared log to ensure that it reproduces the original offset log. Bed boundaries and layer property values may need adjustment to ensure that the squared log reproduces the offset well log. Some of the boundaries between layers are selected as markers (black lines in Fig. 1-10) because they indicate the top of a geological sequence of related layers.

The geological structure is generally derived from surface seismic data and well-to-well log correlation. Faults, formation dip, or changes in layer thicknesses are indicated by changes in the surfaces (blue lines in Fig. 1-10), which represent the tops of geological sequences. The markers in the layer column are associated with surfaces in the geological model and the formation property is propagated into the geological model between the surfaces.

Figure 1-11. There are three property propagation methods.

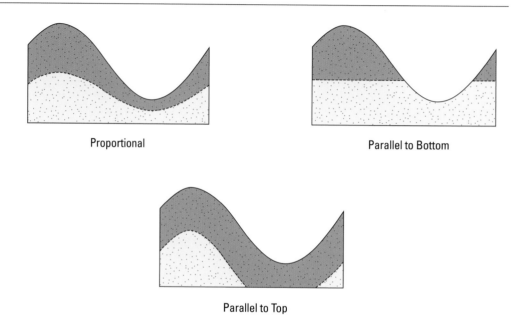

Proportional

Parallel to Bottom

Parallel to Top

The distribution of the properties between the surfaces depends on the geological deposition of the layers. Generally three options are available (Fig. 1-11): proportional propagation, propagation parallel to the upper surface, and propagation parallel to the lower surface. The selection of the propagation method should be based on the way the subsurface layers are actually deposited. Proportional propagation is the most common method.

1.5.1.2 Computing tool responses

Once the geological model has been populated with the formation properties to create a property model, the planned well trajectory is introduced and the properties at each point along the trajectory are derived from the property model. Thin layers, formation boundaries, and the incidence angle between the wellbore and the layers all have an effect on LWD measurements, which can complicate the tool response to a formation property in a particular layer. For this reason the LWD measurement response is computed (forward modeled) at each point along the trajectory based on the thicknesses and properties of the layers surrounding the tool and the incidence angle between the tool (that is, the borehole) and the layers. See Sidebar A for a simple example of forward modeling and its inverse, inversion.

Forward modeling and inversion

Forward modeling is the process of computing a theoretical response given a known set of conditions. In the case of subsurface formation logging the logging tool response is computed based on the formation properties and the wellbore trajectory relative to the structure of the formation.

A simple example of forward modeling involves a tube containing mercury and a pot of boiling water (Fig. A-1). It is known that water boils at 212 degF [100 degC] at standard pressure. If the tube is immersed in the boiling water, the mercury expands and rises up the tube. Knowledge of the thermal expansion coefficient of mercury can be used to forward model how far up the tube the mercury surface rises (that is, calculate the response of the mercury). This forward modeling example shows how a response is predicted from a known set of conditions.

The inverse problem, determining the water temperature based on the response of the mercury, requires calibrating the measurement response to the property of interest (Fig. A-2). In the case of mercury expanding in a tube, temperature graduations can be marked on the outside of the tube. The temperature of the water can simply be read then from the level of the surface of the mercury.

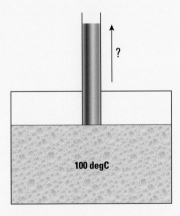

Figure A-1. Forward modeling is used to calculate the height to which mercury rises in a tube (the response) immersed in boiling water (the known conditions).

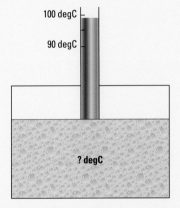

Figure A-2. Inversion uses the expansion of mercury in a tube (the response) to determine the temperature of the water (the unknown condition or property) in which the tube is immersed.

Inversion of logging tool responses is more complicated than the water temperature example. Factors such as vertical and radial volumes of investigation, incidence angle between the wellbore and formation layers, as well as environmental complications such as borehole and invasion effects must be taken into account to derive the true formation properties.

Note that because the geological model is populated with the formation properties and uses the planned well trajectory, the forward-modeled log responses reflect what the real-time logs will look like only if the geology is as mapped and the trajectory is drilled exactly as planned (Fig. 1-12).

To be able to recognize deviations from the plan, alternative scenarios should be modeled. For example, log responses as the well exits the reservoir layer through the roof (top of the layer) and the floor (bottom of the layer) should be modeled so that these situations can be recognized during drilling. Both the structural map and offset well log data should be available for reference in case the well encounters any unexpected features, such as faults or dip changes.

It is important that the operating company's well placement objective is clearly defined and understood, that the formation is modeled as accurately as possible, and that likely alternative scenarios are investigated. During this prejob modeling phase it is the responsibility of the well placement engineer to ensure that the selected LWD measurements are sufficient to achieve the well placement objective. If there is a need for additional measurements (for example, quadrant data or real-time images for formation dip determination), then the value of these measurements and the well placement limitations without them must be recognized early enough to provide sufficient time to make the necessary adjustments to the acquisition program.

Figure 1-13 shows an example of a prejob formation model with the LWD logs forward modeled along the planned trajectory. The three horizontal log tracks at the top of the panel show the modeled GR (green curve), phase resistivity (blue curve), and ring resistivity (black curve). The lower panel shows a vertical section along the planned well trajectory on which the formation GR property model is shown with dark colors representing layers with high GR values and light colors indicating layers with low GR values. The planned well trajectory (green dashed line) targets the middle of the low-GR reservoir layer, indicated by the light colors.

1.5.1.3 Real-time comparing and model updating

After the prejob formation model is prepared, the logs forward modeled along the planned trajectory for the LWD tools, and several alternative scenarios also modeled, the next step is to stream real-time data from the downhole tools and compare the real data with the modeled responses. For details of how the data is transmitted from the downhole tools to the surface in real time, refer to Chapter 4, "Measurement-While-Drilling Fundamentals."

Once the data is transmitted to the surface it must then be distributed to the well placement team. This is typically achieved by uploading the data to a secure, Web-enabled distribution service.

With the data available to the decision makers, it must be visualized and compared with the modeled responses to evaluate any deviations from the plan. Software such as RTGS* Real-Time GeoSteering combines the ability to create and modify a formation model with real-time data streaming and image and log forward modeling along with the ability to display and compare all this information in real time to enhance well placement decision making.

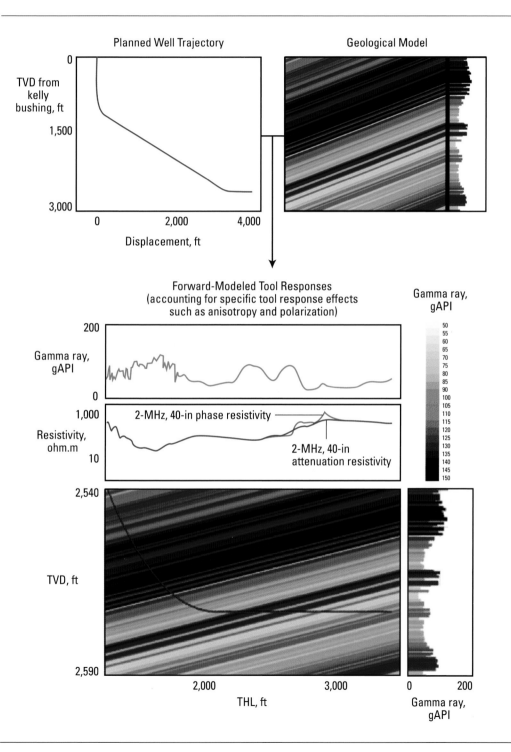

Figure 1-12. The forward-modeled log response along the planned trajectory in the expected formation displays the perfect case only where the geology and well are exactly as expected.

Figure 1-13. At the end of the prejob modeling phase, a structural model populated with the formation properties from an offset well has been built and the theoretical log responses forward modeled for the proposed LWD tools along the planned trajectory.

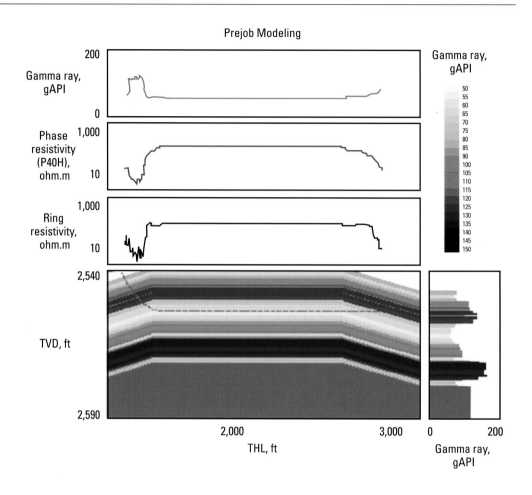

Because the human mind works better with images than numbers, visualization is a powerful technique for simplifying the interpretation of spatially distributed data such as real-time images and quadrant data. Software is available for 2D and 3D visualization and dip picking from LWD images.

The real-time trajectory and log data are streamed into the forward-modeling software usually via a Web data-delivery service. Initially the formation model remains the same, but because the real trajectory may differ from the planned trajectory, the forward-modeled logs are recomputed along the real trajectory. In this stage the measured and modeled logs are compared. If they show a good match, this indicates that the formation model and the location of the real trajectory within in the formation are representative of what is occurring downhole (Fig. 1-14). Where the measured and modeled logs do not agree (Fig. 1-15) suggests that the formation model is not representative of the subsurface and must be updated. The structural model can be changed by adjusting the dip of surfaces or layer thicknesses or by inserting faults.

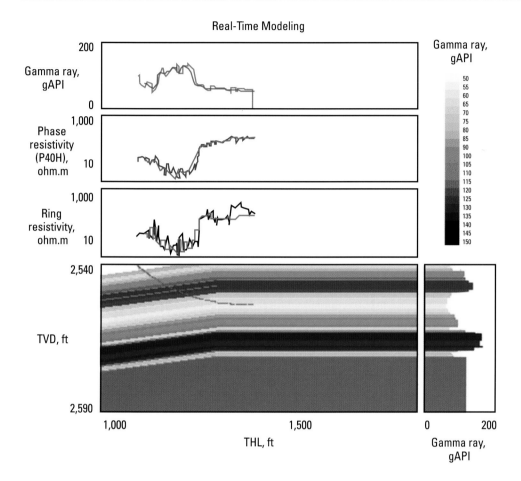

Figure 1-14. During real-time well placement operations, measured data is compared with the modeled logs (red curves). Where they show a good match, as in this example, indicates that the formation model (bottom panel) and the location of the trajectory within the formation are representative of what is occurring downhole.

Typically in well placement using the model, compare, and update technique, the first formation model adjustment is a vertical (TVD) shift to tie in a clear formation feature with the depth at which it is observed in the real well. No further TVD shifts are permitted from this point forward (unless a fault is inserted in the model) because they would disrupt the original tie-in correlation. The convention is to work from left to right across the model in the direction of increasing measured depth. Each new correlation creates either a change in the local dip of the formation or a change in the layer thickness. In the absence of any information about layer thickness changes, the formation dip is generally adjusted.

Figure 1-15. When there are differences between the measured and modeled logs, as shown by the divergence of the log curves from the model (red curves) in the three top tracks, the formation model must be updated to more accurately reflect the subsurface.

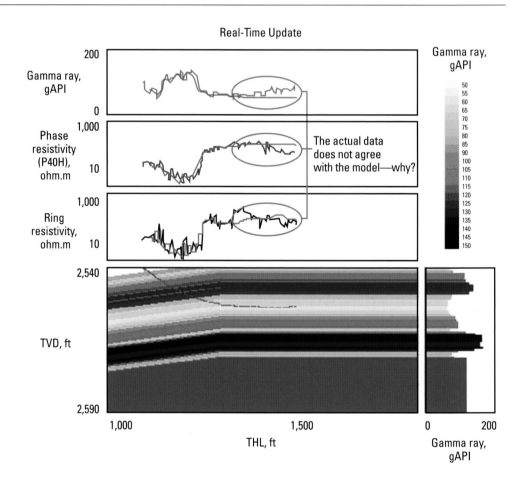

Figure 1-16 demonstrates how a change in formation dip can be used to adjust the match between the measured and modeled logs. In this case, the formation dip was increased slightly so that the real trajectory now lies in the lower part of the lighter colored reservoir section, rather than in the middle as shown in Fig. 1-15. When the log responses are modeled with the well traversing the lower layers in the reservoir, the match between the measured and modeled logs is much improved, indicating that it is a good representation of the well in the subsurface layers. Note that Fig. 1-16 shows the well toward the bottom of the reservoir interval; it should be steered up to stay in the middle of the target reservoir section.

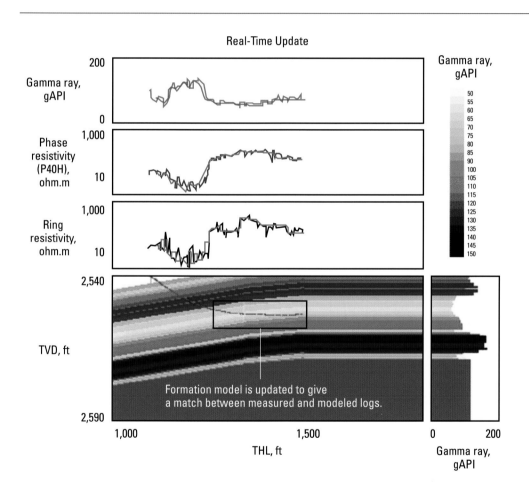

Figure 1-16. Adjusting the modeled formation dip by increasing it slightly from that shown in Fig. 1-15 results in the well trajectory crossing formations that give a better match to the measured logs.

This iterative procedure of modeling the tool responses, comparing them with the real data, and updating the formation model so that they match is repeated as the well is drilled to determine the position of the well trajectory in the layers and make changes to it to ensure that the well stays in the target layer.

Figure 1-17 shows the final well location on the updated reservoir model. Note the additional 400 ft [120 m] of reservoir exposure in comparison with the geometrically drilled well shown in Fig. 1-18.

Figure 1-17. Evaluation of the well location within a reservoir layer enhances both placement of the well in real time and understanding of the formation structure. The modeled curves are red.

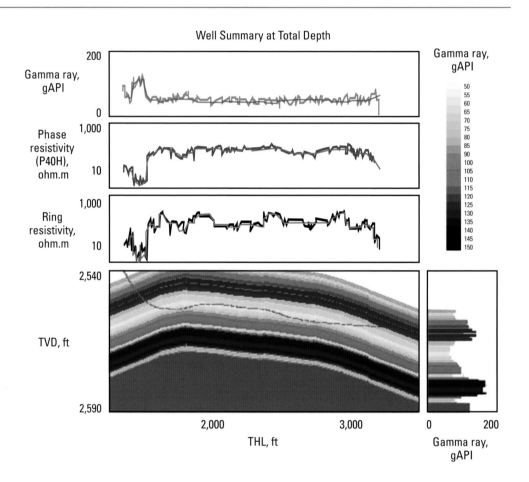

Making appropriate adjustments to the trajectory as it is being drilled can keep the well in the target layer, thereby increasing reservoir exposure and hence production. In addition, knowledge of the structure of the reservoir is improved. This information can be used in refining the 3D reservoir model for subsequent reserves calculations, production simulation, and future well planning.

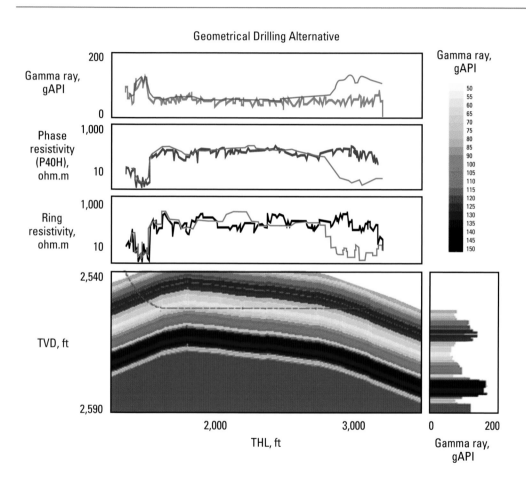

Figure 1-18. Drilling the well geometrically according to the original plan (modeled curves in red) would have delivered less reservoir exposure and production.

1.5.2 Real-time dip determination from azimuthal measurements

1.5.2.1 Limitations of nonazimuthal measurements

The model, compare, and update technique for well placement can be applied to any LWD data. If only average data (that is, data without azimuthal sensitivity) is used, the technique is limited by its inability to distinguish between a boundary that is approaching[†] the well from above versus one approaching the well from below or from a lateral change in the layer being drilled.

Figure 1-19 shows three different scenarios that could explain a decrease in average density porosity, each of which requires a different well placement response.

1.5.2.2 Azimuthal measurements

Azimuthal measurements, which are measurements focused on a sector of the borehole rather than the average around the borehole, provide the information required to distinguish the direction from which features approach the borehole. In contrast to wireline imaging tools that have an array of sensors, LWD imaging is achieved by using BHA rotation to scan a single set of sensors around the inner surface of the borehole. The information is binned into azimuthal sectors. The azimuthal aperture, or width of the sectors, depends on the degree to which the measurement can be focused. Neutron measurements are difficult to focus and hence are generally presented as an azimuthal average. Formation GR measurements can be focused into four quadrants around the borehole, which are generally defined as the bottom, left, up, and right (B-L-U-R) quadrants. For the density measurement, each of the quadrants can be further subdivided into 4 sectors, giving a total of 16 sectors around the borehole in a density image. The currents of the laterolog resistivity measurement can be even more tightly focused into 56 sectors around the borehole.

[†] The drilling perspective is referenced to the wellbore, so layers approach and cross or traverse the wellbore, although in reality the situation is the opposite.

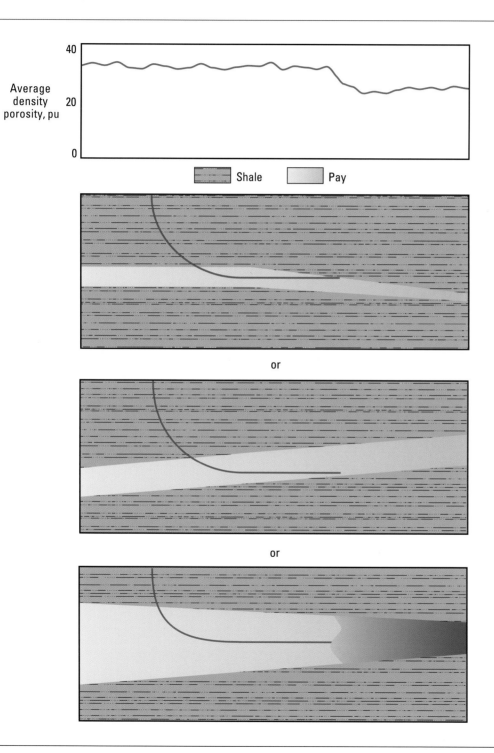

Figure 1-19. Average measurements alone can be ambiguous because they do not distinguish between the well exiting the top or bottom of the reservoir or lateral changes in the reservoir.

Figure 1-20 shows examples of quadrant and sector measurements. Note that the density and photoelectric images are subdivided into 16 sectors, whereas the density and photoelectric measurements are delivered as quadrants. This is because these statistical measurements require counts from four sectors to have the statistical precision required for use as quantitative measurements. Similarly, the laterolog image is delivered with 56 sectors whereas the resistivity measurements are in quadrants because of signal-to-noise considerations.

Figure 1-20. The azimuthal resolution of each measurement depends on how tightly the measurement can be focused. Because quantitative measurements generally require averaging across several sectors to improve measurement signal-to-noise, the image is delivered in sectors but the quantitative measurements are generally delivered as quadrant data.

4 quadrants

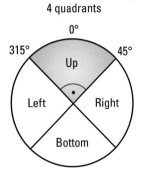

Laterolog resistivity measurement
Density measurement
Photoelectric measurement
GR image and measurement

16 sectors

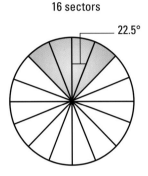

Density image
Photoelectric image

56 sectors

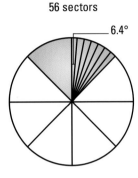

geoVISION* laterolog
resistivity image

The axial resolution of images is related to how tightly the measurement can be focused. Figure 1-21 compares the pixel sizes of various LWD and wireline imaging technologies. The axial resolution of the LWD images is influenced by the drilling ROP compared with the sampling rate of the imaging measurement. For example, when drilling at 36 ft/h [11 m/h], the LWD tool moves 1 ft [0.3 m] in 100 s. If the measurement cycle takes 10 s, then 10 samples per foot, or 1 sample each 1.2 in [3 cm], are acquired, setting the lower limit to the axial resolution.

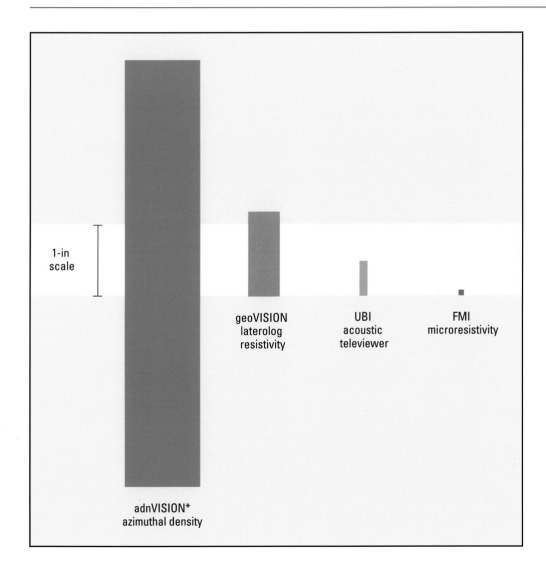

Figure 1-21. Comparison of the relative pixel size of common LWD and wireline imaging technologies in a 6-in [15-cm] borehole. Each pixel represents the area of the borehole wall resolved. From the scale bar on the left: LWD density image pixel, 16 sectors; LWD laterolog image pixel, 56 sectors; wireline UBI* Ultrasonic Borehole Imager pixel; and wireline FMI* Fullbore Formation MicroImager pixel.

The objectives of acquiring the image should be kept in mind when selecting one of the available technologies. If real-time formation dip is required for well placement, then real-time LWD images are the only option. If detailed fracture analysis is the objective, then a high-resolution wireline imager may be required. Compared in Fig. 1-22 are a wireline FMI image (left) and LWD MicroScope* image (right), which are obtained with similar physics of measurement. Although the FMI image shows greater detail, the MicroScope image has full borehole coverage and shows the major formation features required for evaluation of the structure around the well. It also has the significant advantage of being available while drilling.

Figure 1-22. Comparison of dynamically normalized images from the wireline FMI (left) and LWD MicroScope (right) tools shows that the major features seen on the higher resolution FMI image are also visible on the MicroScope image.

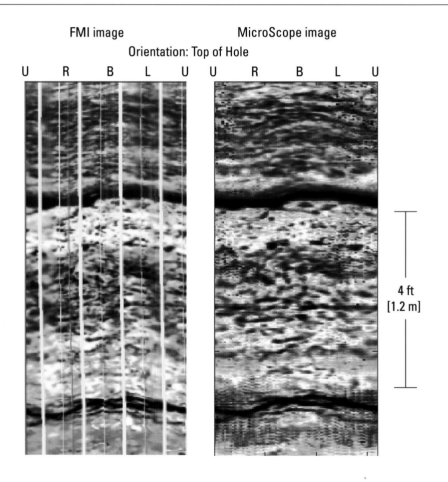

Azimuthal measurements detect changes in formation properties around the borehole. Images made with different measurement physics from the same borehole can show different features because the formation properties they measure do not change in the same way. Figure 1-23 shows five LWD images acquired at the same time in the same borehole. The acoustic image on the far left responds to tool standoff and borehole breakouts; the photoelectric image responds to changes in formation lithology; the density image responds to changes in formation lithology, porosity, and fluid content; the GR image responds to the total formation gamma ray count (from thorium, uranium, and potassium); and the resistivity image responds to the porosity and water content of the pore space. Although the images show common events that can be correlated, the differences between them yield additional information about the formation.

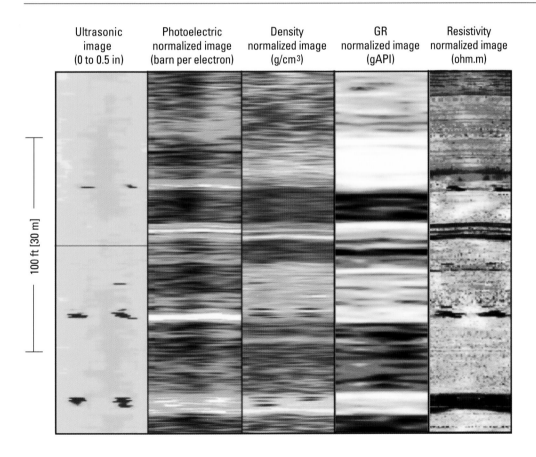

Ultrasonic image (0 to 0.5 in)   Photoelectric normalized image (barn per electron)   Density normalized image (g/cm³)   GR normalized image (gAPI)   Resistivity normalized image (ohm.m)

100 ft [30 m]

Figure 1-23. The various LWD images respond to different formation properties and hence display different formation features.

It is the formation structural information contained in azimuthal data and images that is of interest for well placement. When the borehole crosses a layer with some contrast in formation property relative to the zone being drilled, the azimuthal measurements detect the layer as it is traversed from one side of the borehole to the other. The 2D LWD images are generally displayed as if the borehole had been split along the top and unfolded so that the center of the image corresponds to the bottom of the borehole (Fig. 1-24).

When a borehole is drilled down into a layer, the bottom of the borehole sees the layer first, then the sides see it, and finally the top does. On the image the layer appears first in the middle and then creates a sinusoidal shape that ends at the edges of the image, corresponding to where the last of the layer is seen at the top of the borehole. The amplitude of this sinusoid is related to the incidence angle between the wellbore and the layer. Even without calculation of the incidence angle, the "happy face" and "sad face" features on an image that result respectively from drilling up through layers and down through layers can assist in well placement decision making because they indicate where the wellbore is positioned relative to the layering. For example, if happy face features are seen on the image while drilling in a reservoir, this indicates drilling up through the sequence, and a drop in well inclination may be required to become parallel to the layering and remain in the reservoir.

Because it is responding to the local geometry of the layering with respect to the tool, a happy face feature on a borehole image always indicates that the well is penetrating up into a layer. Similarly, a sad face feature always indicates that the well is drilling down into the top of a layer, irrespective of the trajectory, layer dip, or azimuth.

As the incidence angle between the borehole and layer decreases, the amplitude of the sinusoid on the image increases. This means that in a high-angle well through horizontal layering, a thin bed appears stretched over a significantly longer MD than in a nearly vertical well.

For example, in an 8.5-in [21.6-cm] well drilled at an incidence angle of 1° through a 6-in- [15-cm-] thick bed, the bed appears on the image stretched over more than 69.2 ft [21.1 m] of MD. Refer to Fig. 5-4 for a sketch of this geometry. In a well drilled perpendicular to the layer, the 6-in-thick bed would not be fully resolved by a measurement with an axial resolution of 1 ft [30 cm]. However, in a well close to parallel with the bed, even a 1-ft axial resolution measurement is able to identify the feature because it is stretched over 69.2 ft.

This geometrical stretching of formation features in high-angle wells greatly enhances the visibility of thin layers. Consequently even layers that are too thin to be fully identified in nearly vertical wells can be used for well placement purposes in high-angle wells.

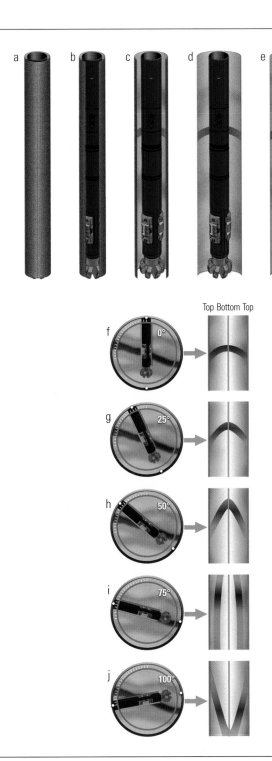

Figure 1-24. a through e: The 2D presentation of azimuthal LWD data splits the borehole along the top and unfolds it with the middle of the image corresponding to the bottom of the borehole. f through j: Tool inclination is adjusted from vertical, through horizontal, to drilling up at 100°. The bedding plane is parallel to the wellbore at an inclination angle of about 75°. Parts f, g, and h reflect drilling down section (the layer being crossed by the borehole creates a "sad face" on the image). Part i shows the parallel "railroad tracks" characteristic of drilling at the same angle as the bedding plane. Part j shows the "happy face" that occurs when the borehole drills up through a layer. In each case the amplitude of the sinusoid is characteristic of the incidence angle between the borehole and the layer.

1.5.2.3 Image color scaling

Color is used to encode information about the magnitude of the formation parameter used in the creation of the image. For example, a density image is acquired by a single set of density detectors (GR source, long- and short-spacing scintillating detectors) as they scan the internal surface of the borehole during rotation of the drillstring. The density of the formation is independently measured in each of 16 sectors (360°/16 sectors = 22.5° of borehole azimuth coverage per sector). To convert this density information into an image, the density value is mapped to a color, typically with darker colors indicating lower density whereas lighter colors indicate higher density.

Image normalization is a method by which features can be visually enhanced (Fig. 1-25). A static image has a fixed, user-defined scale over which the color spectrum is applied. For example, the colors of a density image may be scaled over 16 colors, with dark colors starting at 2.2 g/cm^3 to light colors at 2.7 g/cm^3. A dynamic image uses a depth window within which the minimum and maximum values are used as the endpoints for the color scaling. This method allows subtle contrasts between features to be enhanced.

It is generally good practice to use both static and dynamic images because the static image highlights large-scale features whereas the dynamic image enhances features within the normalization depth window, which can reduce the visibility of the large-scale features. The dynamic image should never be used for feature identification without reference to the static image to confirm the presence of the feature. This is because in horizontal wells where there is a relatively small change in formation properties the dynamic normalization processing tends to enhance noise such as that caused by wellbore rugosity.

Figure 1-25. In these statically (upper) and dynamically (lower) normalized images over the same interval, the static image has a fixed density range for each color. The dynamic image uses the full range of available colors over a user-specified depth window.

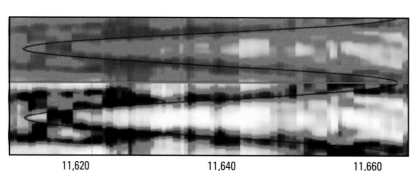

Statically normalized image

Dynamically normalized image

11,620　　　　11,640　　　　11,660

MD, ft

1.5.2.4 Image acquisition considerations

The acquisition of azimuthally focused formation data, including images, requires that the BHA is rotating so that the LWD sensors are scanning the circumference of the borehole. The orientation of the sensors relative to the top of the hole (in high-angle and horizontal wells or the north reference in nearly vertical wells) must be known for each acquisition sample to allocate the data to the appropriate azimuthal sector. This orientation information is acquired using two orthogonal magnetometers, which provide sinusoidal responses as they rotate in the Earth's magnetic field (Fig. 1-26).

Magnetometers are used rather than accelerometers (which would enable direct determination of the top of hole in high-angle and horizontal wells) because the accelerations associated with BHA rotation and shocks would induce noise on the accelerometer-derived sinusoids, making it difficult to accurately determine the orientation to the top of the hole. The magnetic environment is quieter. However, because the Earth's magnetic and gravitational fields generally do not coincide, azimuthal data oriented to the magnetic field must be rotated to orient it to the gravitational field, also called the top-of-hole reference.

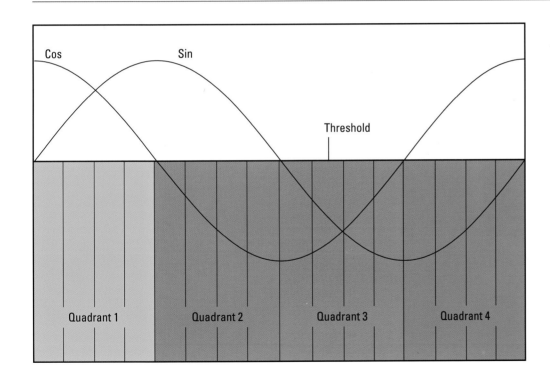

Figure 1-26. The sinusoidal responses of two orthogonally mounted magnetometers are used to determine the sensor orientation with respect to the Earth's magnetic field.

During rotation of the LWD tool about its axis, the sensors sweep out a circle in the plane perpendicular to the axis of the tool. The orthogonal magnetometers, mounted perpendicular to the axis of the tool, measure the projection of the Earth's magnetic vector in this plane. The projection of the Earth's gravitational vector in this plane defines the direction of the bottom of the hole. Data acquired relative to the projection of the magnetic vector must be rotated through the angle shown in red on Fig. 1-27 to orient it relative to the projection of the Earth's magnetic vector.

In a vertical well there is no projection of the Earth's gravitational vector in the plane perpendicular to the axis of the tool. In this case azimuthal data must be oriented to a north reference because a top-of-hole reference is meaningless in a vertical well.

Because the orientation of azimuthal data requires a projection of the magnetic field on the plane perpendicular to the axis of the tool, care must be taken when drilling close to parallel to the Earth's magnetic vector.

Figure 1-27. Azimuthal data is initially oriented relative to the projection of the Earth's magnetic vector on the plane perpendicular to the axis of the tool. The data must be rotated to orient it to the top-of-hole reference, which is the standard orientation in high-angle and horizontal wells.

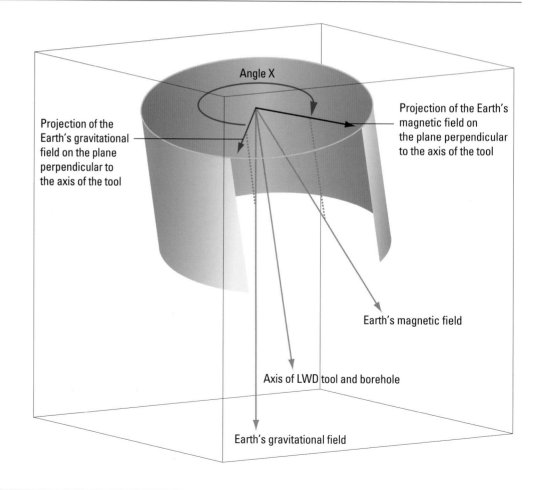

Angle X

Projection of the Earth's gravitational field on the plane perpendicular to the axis of the tool

Projection of the Earth's magnetic field on the plane perpendicular to the axis of the tool

Earth's magnetic field

Axis of LWD tool and borehole

Earth's gravitational field

If the well is drilled along the magnetic vector there is no projection of the magnetic field detected by the orthogonal magnetometers in the tool (Fig. 1-28). In this case azimuthal data cannot be oriented to either a north reference or the top of hole.

This zone of exclusion occurs when both of the following conditions exist:

- well azimuth within 5° of being parallel to the Earth's magnetic field

- well inclination within 5° to 10° of being parallel to the Earth's magnetic field.

Within this cone around the Earth's magnetic field, the amplitude of the magnetic field sinusoids detected by the orthogonal magnetometers remains too small to reliably orient azimuthal data. During well planning a zone of exclusion is defined around the Earth's magnetic vector in which azimuthal data cannot be acquired. Wells are generally planned with the minimum possible interval in the zone of exclusion.

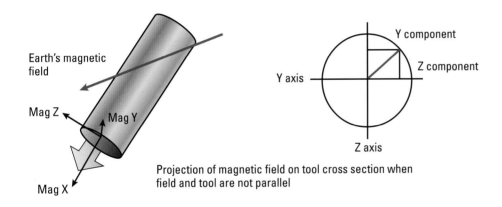

Projection of magnetic field on tool cross section when field and tool are not parallel

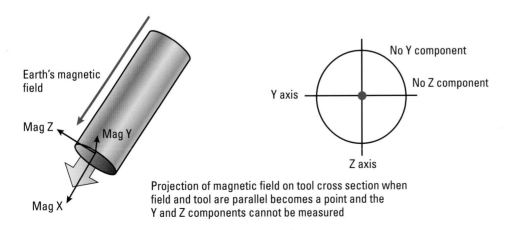

Projection of magnetic field on tool cross section when field and tool are parallel becomes a point and the Y and Z components cannot be measured

Figure 1-28. A zone of exclusion exists around the Earth's magnetic vector where the tool orientation within the borehole cannot be determined.

1.5.2.5 Using images to calculate the incidence angle between the borehole and a layer

The incidence angle between the borehole and a formation layer defines a triangle.

- Adjacent side—The amplitude of the sinusoid on the image is measured along the borehole. The MD amplitude of the sinusoid must be converted to the same units as the borehole diameter.

- Opposite side—Images do not scan the surface of the borehole. They represent the formation property at the depth of investigation of the measurement from which they are derived. Therefore, the depth of investigation of the imaging measurement must be added on each side of the borehole diameter. For density measurement, the depth of investigation is approximately 1 in [2.5 cm], so for an 8.5-in [21.6-cm] borehole the diameter at which the image is acquired is 10.5 in [26.7 cm]. For laterolog resistivity images the electrical penetration depth is approximately 1.5 in [3.8 cm], so 3 in [7.6 cm] should be added to the diameter of the borehole to calculate formation dip from an LWD resistivity image.

Figure 1-29 shows a wellbore crossing a layer with the corresponding density image and incidence angle calculation.

Figure 1-29. The incidence angle between the borehole and formation layering can be deduced by solving the trigonometry of a triangle. DOI = depth of investigation.

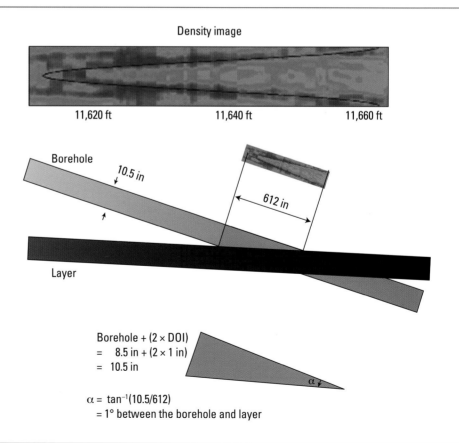

Density image

11,620 ft   11,640 ft   11,660 ft

Borehole

10.5 in

612 in

Layer

Borehole + (2 × DOI)
= 8.5 in + (2 × 1 in)
= 10.5 in

α = tan^{-1}(10.5/612)
= 1° between the borehole and layer

1.5.2.6 Strike, true dip, and azimuth

For the computation of dip, formation layers are assumed to be planar surfaces (Fig. 1-30). The strike of the plane is the azimuth of the intersection of the plane with a horizontal surface. In other words, the strike of the plane is the azimuth in which there is zero dip. Dip is the magnitude of the inclination of a plane from horizontal. The true, or maximum, dip is perpendicular to the strike.

The plane in Fig. 1-30 has a strike northeast–southwest (45°) and a dip of 30° at an azimuth of 135° (southeast).

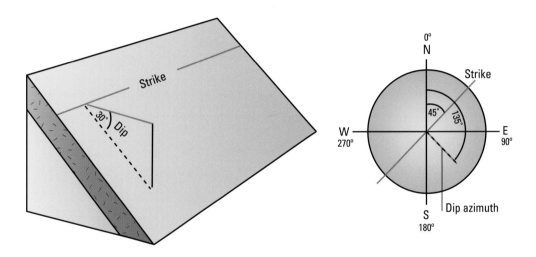

Figure 1-30. Strike is the azimuth of the intersection of a plane, such as a dipping bed, with a horizontal surface. Dip is the magnitude of the inclination of a plane from horizontal. True, or maximum, dip is measured perpendicular to strike.

1.5.2.7 Apparent dip and incidence angle

If the azimuth of the formation dip and the azimuth of the borehole inclination are different, then the apparent angle to the formation layering as measured in the azimuth of the wellbore is less than the true formation dip (Fig. 1-31). A horizontal well being drilled along the strike line (A in Fig. 1-31) sees the apparent dip of the layer along the length of the borehole as zero. A well drilled perpendicular to strike sees the true dip of the layer. A well drilled at any other azimuth, such as along line A-B, sees an apparent dip of the layer that is less than the true dip.

Figure 1-31. The projection of formation dip on a vertical plane is at a maximum when the vertical plane is oriented perpendicular to the strike, giving true dip. Projection in a vertical plane along any other azimuth has a lower dip value, the apparent dip.

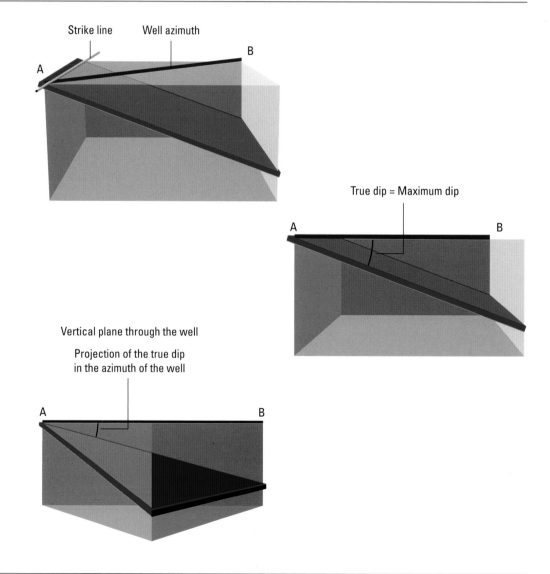

Apparent dip refers to the projection of the true dip in an azimuth other than perpendicular to strike. In a well placement context, the more significant angle is the incidence angle between the borehole and the layer (Fig. 1-32).

To remain parallel to a layer while drilling, the incidence angle is all that is required to make trajectory change decisions, provided that the well continues to be drilled in the same azimuth. If the well changes azimuth (that is, turns left or right) then the incidence angle will change, and this must be taken into account to be able to stay in the target layer.

Calculating the true dip of the formation from the incidence angle requires that the borehole trajectory be taken into account. This is because the incidence angle is measured by the logging instruments from the perspective of the borehole. For example, the incidence angle between a vertical borehole and formation layers dipping at 45° is the same as the incidence angle between a 45° inclined borehole and horizontal layers.

1.5.2.8 Image orientation for formation dip determination

Converting from incidence angle to true dip must account for the inclination and azimuth of the borehole. Many static surveys of borehole azimuth are referenced to grid north instead of true north, specifically so the directional driller can determine from the plan view of the survey whether the trajectory is "on plan" and within the lease boundaries. However, because geology data is generally referenced to true north, the grid correction must be removed before using the MWD survey to interpret LWD images. The accuracy of MWD surveys is typically ±1° in azimuth, so the grid correction should be removed in most cases because it can be in the order of a few degrees. Errors in wellbore azimuth translate into errors in formation dip when transforming incidence angle to true dip. For details on north references, refer to Section 4.3.3, "Azimuth."

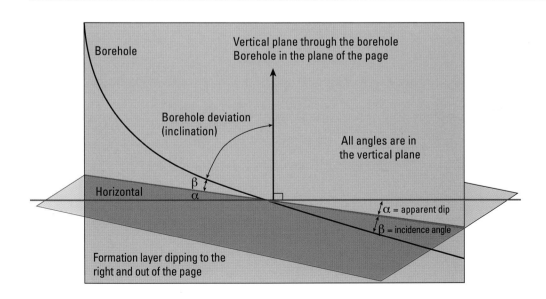

Figure 1-32. The incidence angle is measured between the wellbore and the formation layer. Apparent dip is the angle between the formation and the horizontal.

1.5.2.9 Visualization in 3D

Real-time 3D visualization and dip-picking software is available to assist in visualizing the intersection of formation layers and inclined boreholes. By wrapping an oriented image around a 3D representation of the corresponding well trajectory, 3D visualization software helps the user understand the subsurface orientation of the wellbore and layers that it intersects.

Figure 1-33 shows a WellEYE* 3D borehole data viewer display. The right panel is a conventional 2D display of laterolog resistivity log and image data. The green sinusoid on the image tracks is where a geological feature has been picked. The corresponding formation dip information is indicated by the green tadpole in the dip track between the two image tracks. The left panel shows a 3D perspective of the well trajectory with the laterolog image data "wrapped" around the borehole. The dip picked on the 2D image is represented by the green plane crossing the 3D trajectory, visualizing the intersection of the borehole and layer. On the 3D display the MD of the well increases as the well gets closer, indicating that the well has just been drilled up through the bottom of the layer indicated by the green plane. If the objective is to stay in the higher resistivity (lighter color) layer below the green plane, then the visualization clearly shows that the well should be steered back down into the target layer.

The ability to transmit real-time image data from downhole and stream it into interactive visualization and dip-picking software, either at the wellsite or through an Internet connection, enhances the value of the information because it enables almost instantaneous quantitative interpretation to improve decision making.

1.5.2.10 Using quadrant data to calculate incidence angle

Although images make the recognition and correlation of an event across the borehole easier, a similar incidence angle calculation can be performed based on quadrant data alone if correlation of the features in the up and bottom quadrants is clear. The same logic applies for the calculation of incidence angle, although rather than using the amplitude of the sinusoid, the difference in MD between correlated events in the up and bottom quadrants is used.

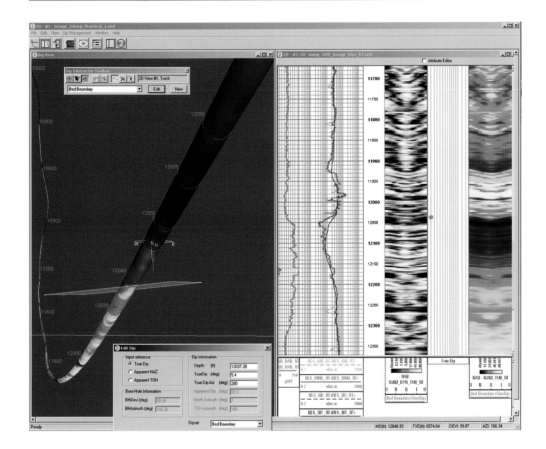

Figure 1-33. Real-time 3D visualization and dip-picking software enhances the user's understanding of the subsurface geometry.

The example in Fig. 1-34 is a typical LWD log from a horizontal borehole. The top track shows quadrant density responses and an azimuthal average thermal neutron response. The middle track shows three phase resistivities of differing depths of investigation. The depth track includes a pink line labeled ARPM (adnVISION revolutions per minute) that indicates where the density tool is rotating. Where this curve drops to zero it indicates that the density tool is not rotating and hence the azimuthal data (which requires rotation to scan around the borehole) is not available. Note that across the interval of no rotation the two quadrant density curves collapse to a single value. Care must be taken to conduct quality control of azimuthal data by checking that the tool was rotating over the interval of interest. The bottom track of Fig. 1-34 shows the GR (green curve), drilling ROP (black dashed curve), and TVD (blue curve) of the borehole.

There is a 4-m [13-ft] MD difference between the density drop seen by the bottom and up density measurements. The bottom density measurement decreases first, which indicates that the well is being drilled down into a layer of lower density. The TVD curve shows that the well is horizontal, so the layer must dip upward.

For this 6-in- [15.24-cm-] diameter well, the calculation of incidence angle (α) is as follows:

MD difference = 4 m = 157.5 in
Borehole diameter = 6 in
Density measurement depth of investigation = 1 in

$$\tan \alpha = (borehole\ diameter + [2 \times measurement\ depth\ of\ investigation])/MD\ difference \qquad (1\text{-}1)$$
$$= (6 + [2 \times 1])/157.5$$
$$= 0.0508$$

$$\alpha = \tan^{-1} 0.0508$$
$$= 2.9° \text{ incidence angle between the borehole and layer.}$$

The incidence angle calculation, derived either from images or quadrant curves, is used to position the wellbore relative to the layer. For the example in Fig. 1-34, if the intention is to maintain the wellbore parallel to the layer, the well inclination must be increased by 2.9° to achieve the well placement objective.

1.5.2.11 Using dip to calculate layer thickness

Fully defining a formation layer requires knowing its true thickness and dip. When directionally drilling through a layer of unknown dip, nonazimuthal logs can provide only the MD thickness of the layer, which can be converted to the true vertical depth thickness (TVDT) based on the well deviation. To define the true bed thickness (TBT), the formation dip must be known (Fig. 1-35).

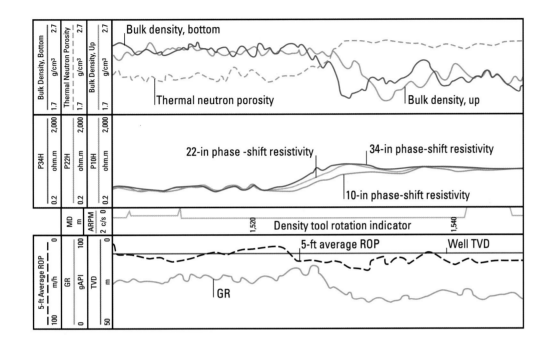

Figure 1-34. This typical LWD log from a horizontal borehole displays quadrant density information in the top track. The bottom density (red curve) and up density (green curve) show a clearly correlated drop in density.

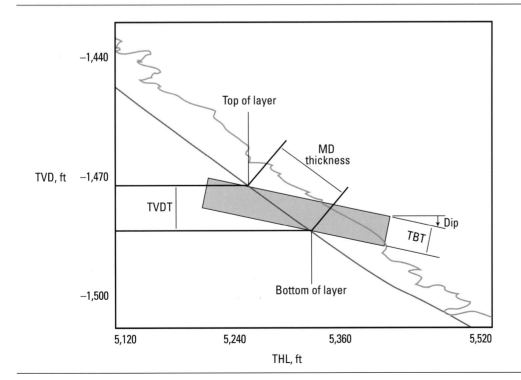

Figure 1-35. The true bed thickness cannot be determined from conventional logs unless the formation dip is known.

In the absence of formation dip information, the conventional model, compare, and update well placement method generally assumes that a formation layer thickness is the same as observed in an offset well (Fig. 1-36).

Figure 1-36. Formation dip can be calculated by assuming that the layer thickness in the drilled well is the same as in an offset well.

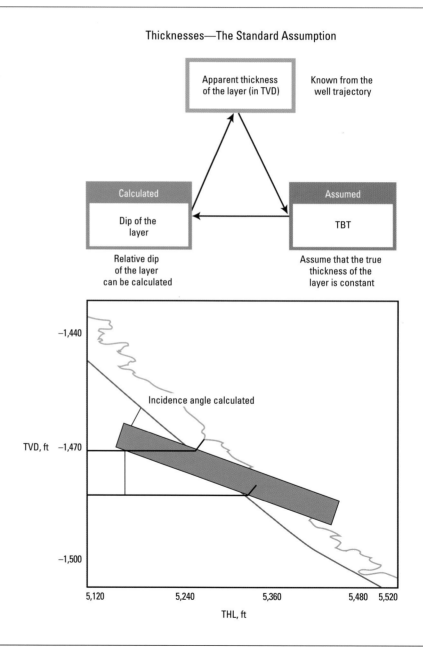

The assumption of constant layer thickness can result in an invalid formation model and poor well placement. Nonazimuthal data cannot distinguish between change in the dip of the layer, change in the thickness of the layer, or change in both. A combination of layer thickness and dip can explain any apparent MD layer thickness (Fig. 1-37).

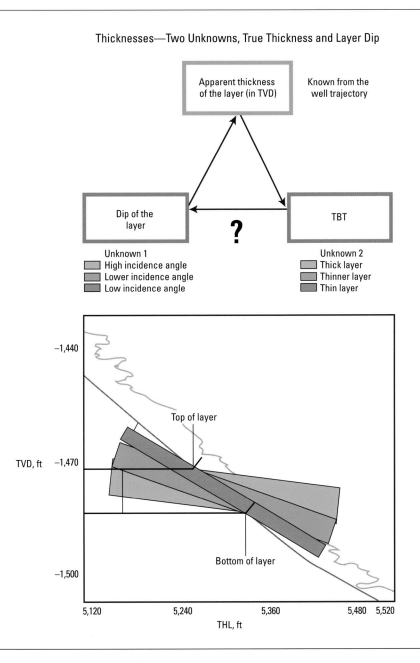

Figure 1-37. Various combinations of layer thickness and dip can explain the same apparent thickness in MD.

By enabling determination of the formation dip at the point where it crosses the borehole, azimuthal data makes it possible to calculate the TBT (Fig. 1-38).

Figure 1-38. Information on the layer dip that is available from azimuthal measurements enables building a more accurate formation model by removing the need to make assumptions.

Thicknesses—Formation Dip Allows True Thickness Calculation

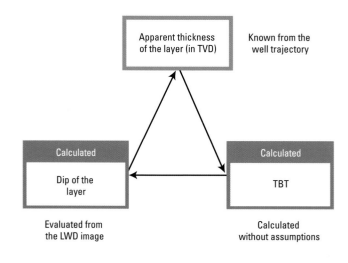

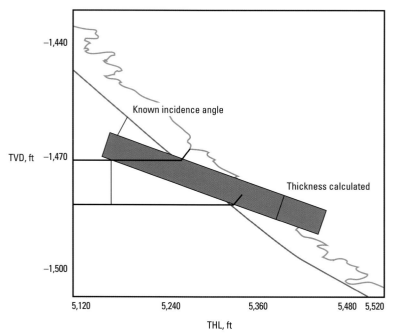

Real-time dip determination from azimuthal data is a powerful, complementary method for use with the basic model, compare, and update technique for well placement. Azimuthal data provides information about the direction from which a feature approaches and crosses the borehole and the dip of the feature, which allows the TBT to be calculated.

1.5.2.12 Projecting to the bit

During real-time acquisition the sensor measure point is normally some distance behind the bit. Azimuthal measurements are used to determine the incidence angle between the formation and borehole, and this information can be used to estimate the position of the bit relative to the layer (Fig. 1-39):

$$d = l\sin\alpha, \tag{1-2}$$

where

d = distance to the trajectory perpendicular to the layer (assumes that the layer and trajectory do not change in angle)

l = length between the azimuthal measure point and the bit

α = incidence angle between the trajectory and layer.

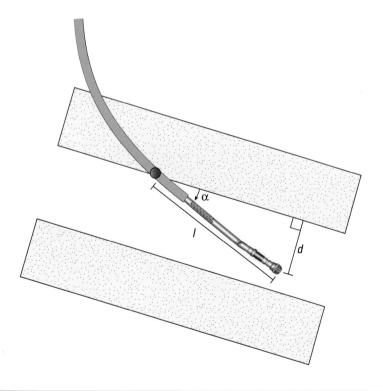

Figure 1-39. An azimuthal measurement is usually located a distance (*l*) behind the bit.

1.5.3 Remote detection of boundaries

Although azimuthal data greatly enhances well placement by enabling determination of the direction from which features contact the borehole, most azimuthal measurements have a limited depth of investigation (of the order of inches). This effectively means that with these measurements a boundary can be detected only when the borehole has come into contact with it.

The development of deep directional electromagnetic measurements has revolutionized well placement by enabling the remote detection of resistivity changes within a formation. The PeriScope bed boundary mapping service complements the LWD measurement portfolio by adding the azimuthal sensitivity previously available only with images while extending the depth of investigation beyond that of conventional LWD propagation resistivity.

Tilted antennae on the PeriScope tool create directional sensitivity to conductivity changes in the formation. Directional measurements of electromagnetic phase shift and attenuation are analyzed downhole to determine the direction to the nearest conductivity contrast and then transmitted from the downhole tool to the surface, where inversion processing extracts the distance-to-boundary (DTB) information.

Figure 1-40. PeriScope inversion processing solves for the resistivity and distance to layer boundaries above and below the wellbore based on a three-layer model (left). The resulting information is displayed in real time as a color-coded resistivity cross section of the formation (right).

R_u = resistivity of the upper layer
R_h = horizontal resistivity of the local layer
R_v = vertical resistivity of the local layer
R_d = resistivity of the lower layer
h_u = distance to the upper layer
h_d = distance to the lower layer

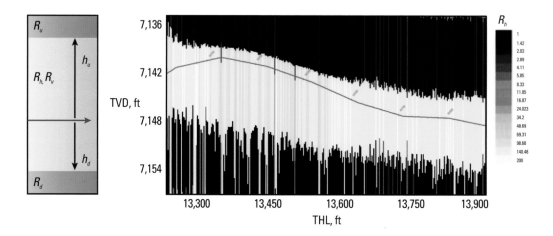

Inversion processing converts raw directional phase-shift and attenuation measurements, acquired at multiple frequencies and transmitter-receiver spacings, into a three-layer formation model. The inversion solves for the resistivity of and distance to the layer above, the resistivity of and distance to the layer below, and the resistivity of the layer in which the well is being drilled. This information is then displayed in real time as a color-coded resistivity cross section of the formation along the wellbore (Fig. 1-40).

In addition to DTB information, the PeriScope service provides an azimuth to the boundaries, based on the assumption that the layers above and below are parallel. This information is presented in an azimuthal view (Fig. 1-41).

The distance and azimuth information enable steering of a well in both TVD and azimuth relative to a resistivity boundary without having to come in contact with it. For example, in the situation shown in Fig. 1-41, the well could be either turned up to avoid the lower boundary or turned to the left, or a combination of the two could be used depending on the most appropriate position for the wellbore in the target layer.

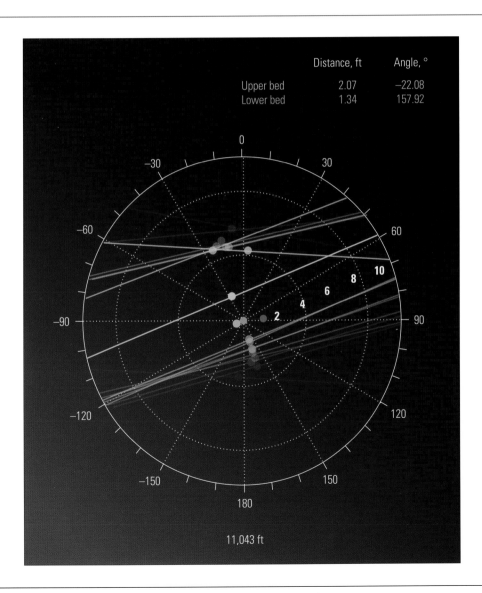

	Distance, ft	Angle, °
Upper bed	2.07	−22.08
Lower bed	1.34	157.92

11,043 ft

Figure 1-41. The azimuthal viewer provides a representation of the subsurface as if looking down the borehole (center) with the upper (yellow) and lower (blue) boundaries shown at the distance and azimuth around the borehole determined from PeriScope measurements. The 10 most recent upper and lower boundaries are shown with the most recent pair in the brightest colors. This display allows evaluation of the DTB trend along the trajectory. The red dot represents the position of the bit when looking along the wellbore.

Seismic data can visualize large structural features, but owing to the limited frequency content of surface seismic data, many features are below seismic resolution. The PeriScope service, with a radius of investigation of approximately 15 ft [4.5 m], can be used to evaluate events such as subseismic faults (Fig. 1-42). As thinner reservoirs are drilled, the evaluation of subseismic faulting becomes increasingly important because the borehole can exit the reservoir on encountering a fault. In addition, the evaluation of reservoir compartments is improved through the ability to track the top and bottom of the layer, thereby improving estimation of the volume of hydrocarbons in place. Further details on the PeriScope service are available in Section 5.14, "Remote boundary detection."

Figure 1-42. PeriScope remote detection of boundaries delivers both directionality and depth of investigation, allowing well placement relative to subseismic features in the reservoir (left). In many cases this is not possible using non-azimuthal measurements such as propagation resistivity (center right), or shallow directional measurements such as those used for borehole imaging (right).

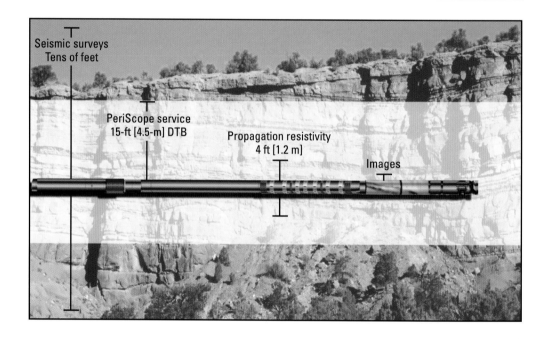

1.6 The three components of well placement

1.6.1 Downhole tools

1.6.1.1 Directional drilling technology

Directional drilling is the technique of deviating a wellbore along a planned course to a subsurface target that has a location at a given lateral distance and direction from vertical.

The drilled wellbore must

- be drilled as safely as possible
- be placed in the required target
- facilitate the planned logging program
- enable smooth running of casing and completion hardware
- not result in excessive casing wear from subsequent operations
- be accessible for future well intervention
- be drilled at the lowest possible cost.

There are a number of directional drilling technologies, such as jetting, whipstocks, and rotary steerable assemblies. The two most commonly used for well placement are the steerable motor and rotary steerable system (RSS).

Steerable motors consist of a positive displacement motor (PDM) with a surface-adjustable bent housing, which enables orienting the bit in the desired drilling direction. PDMs convert hydraulic power from the mud circulation into mechanical power in the form of bit rotation (Fig. 1-43). This is achieved with a progressing cavity design in which the movement of the mud pushes on a rotor with one less lobe than the stator in which it is housed. The rotor turns inside the stator and thus turns the bit coupled to the rotor. With a PDM the bit rotates when there is mud circulation, even if the drillstring is stationary.

By orienting the surface-adjustable bent housing and rotating the bit by pumping mud through the PDM it is possible to drill in a desired direction (Fig. 1-44). The process of keeping the drillstring oriented in a desired direction while drilling is called sliding. The orientation, or toolface, is referenced to the high side of the hole in deviated and horizontal wells. A toolface of 0° indicates building angle, a toolface of 90° indicates drilling right, a toolface of 180° indicates dropping well angle, and a toolface of 270° indicates drilling left. In nearly vertical holes the toolface is oriented relative to north, which is called a magnetic toolface (MTF) to distinguish it from the gravity toolface (GTF) used in high-angle wells.

Because the angle on the bent housing is small (typically less than 3°) the steerable motor can also be rotated. This negates the effect of the bend and gives a relatively straight borehole, which is slightly over gauge (larger than bit size). By alternating sliding and rotating intervals the directional driller can control the rate at which the borehole angle is changed. The rate of change in borehole angle is normally given in degrees per 100 ft [30 m] and is called dogleg severity (DLS).

The difficulty with slide-rotate sequences is that orienting prior to each slide section is time consuming and hence reduces the overall ROP achieved for the well. In addition, because the mud is not agitated during sliding, hole cleaning efficiency is reduced. Finally, the overall length of the well can be limited because static friction during sliding, which is greater than dynamic friction while rotating, can prevent effective weight transfer to the bit. To overcome these limitations, the RSS was developed.

Rotary steerable systems deliver continuous steering while rotating. This capability beneficially results in steadier deviation control, a smoother hole, better hole cleaning, extended hole reach, and overall improvement in the ROP compared with steerable motor performance. It is also advantageous for well placement because the continuous rotation of the BHA ensures that when formation images are acquired, they are available over the entire length of the well.

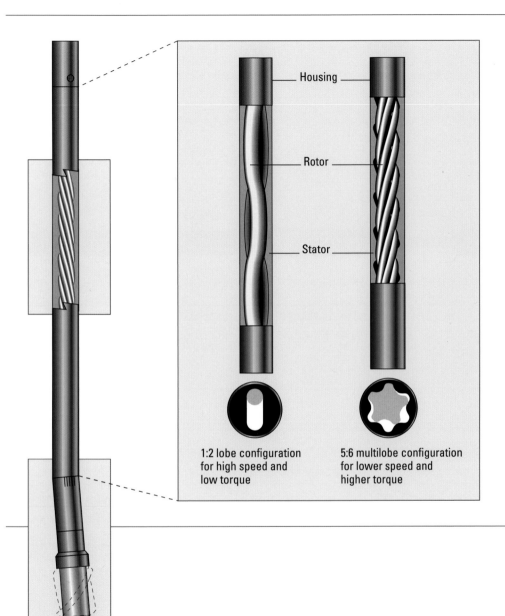

Housing

Rotor

Stator

1:2 lobe configuration
for high speed and
low torque

5:6 multilobe configuration
for lower speed and
higher torque

Figure 1-43. A positive displacement motor converts hydraulic power from the mud flow into mechanical power in the form of bit rotation. Increasing the number of rotor and stator lobes increases the torque available but decreases the speed of bit rotation.

Figure 1-44. The surface-adjustable bent housing is oriented in the direction of the desired well trajectory.

The two main types of RSS are

- push-the-bit—applies side force to increase the side-cutting action of the bit

- point-the-bit—introduces an offset to the drilling trajectory similar to a bent housing but allowing continuous rotation.

A push-the-bit system uses pads on a bias unit to push against the borehole wall, which pushes the bit in the opposite direction (Fig. 1-45).

Figure 1-45. A push-the-bit system uses three pads to push against the borehole wall and so deflect the well in the opposite direction.

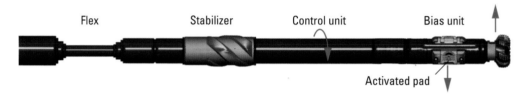

Because the system is rotating, the pads must be activated in sequence to ensure consistent steering in the desired direction. The control unit contains the electronics for control of the toolface and the percentage of time spent steering. The system operates by diverting a small portion of the mud flow to activate the pads. By sensing the rotation of the BHA relative to the Earth's magnetic field and controlling a motor to rotate in the opposite direction, the control unit holds a control valve (blue element in Fig. 1-46) geostationary. The diverted mud flow is directed behind each of the pads in sequence as the entry port to the piston behind each pad rotates in front of the geostationary port in the control valve (Fig. 1-47).

To meet the high power requirements of the control motor, the system generates its own power by using a high-power turbine and alternator assembly above the control unit.

The dogleg delivered by the push-the-bit system depends both on the interaction of the pads with the formation (for example, DLS is lower in unconsolidated formations) and the proportion of time spent steering. The directional driller communicates with the tool through a sequence of mud flow rate changes called downlinking. This method enables the driller to command the tool what proportion of time the control valve should be held oriented (that is, steering) versus time in neutral mode, when the control valve is rotated so there is no active steering.

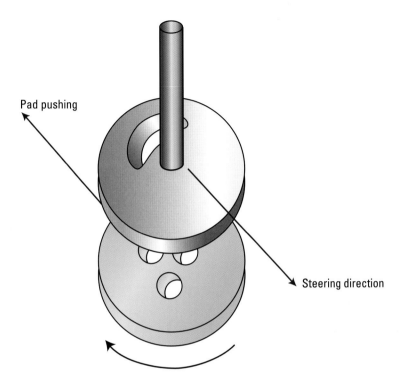

Pad pushing

Steering direction

Figure 1-46. The spindle (blue) is held geostationary by the control unit, thereby diverting a small proportion of the mud flow behind each of the pads in sequence as their respective entry ports rotate in front of the port in the control valve.

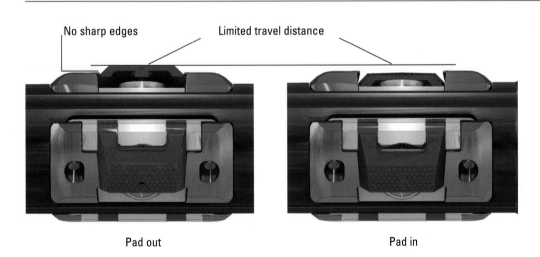

No sharp edges

Limited travel distance

Pad out

Pad in

Figure 1-47. The diameter of the bias unit at the pads is only slightly smaller than the bit size so the pads do not have to travel far before contacting and applying force to the borehole wall. Pad travel is limited to approximately 0.75 in [2 cm].

Compared with the slide-rotate sequences of a motor with a bent housing, the continuous steering of an RSS delivers smoother trajectories, which improves the drilling operation itself, as well as subsequent casing and completion runs and later well intervention operations (Fig. 1-48).

A point-the-bit system delivers the benefits of a push-the-bit system with reduced sensitivity to the formation, resulting in more consistent steering and generally higher dogleg capability.

Figure 1-48. The trajectory oscillations resulting from steerable motor slide-rotate sequences (above) are eliminated by the continuous steering of an RSS (below).

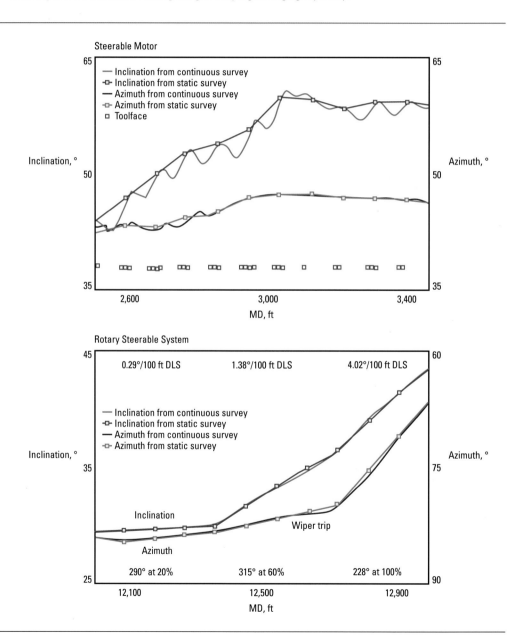

The system is centered on a universal joint that transmits torque and weight on bit, but allows the axis of the bit to be offset with the axis of the tool. The axis of the bit is kept offset by a mandrel that is maintained in a geostationary orientation by a counter-rotating electrical motor (Figs. 1-49 and 1-50). To meet its high power requirements, the system generates its own power with a high-power turbine and alternator assembly. The system also contains high-power electronics to control the motor and sensors that monitor the rotation of the collar and motor. The sensors provide input and feedback for control of the system.

As with the push-the-bit system, the directional driller controls the dogleg by downlinking to the tool to change the proportion of steering versus neutral time.

Because there is no rotation provided downhole by either type of RSS, the entire drillstring must be continuously rotated from surface. If additional downhole rotation is desired or surface rotation must be kept to a minimum (such as when casing wear is a concern), a mud motor (without the bent housing) can be used above the RSS to provide downhole rotation of the RSS assembly.

Further information on directional drilling is in Chapter 3, "Directional Drilling Fundamentals."

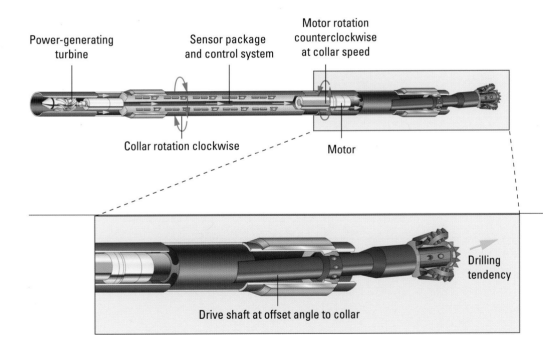

Power-generating turbine

Sensor package and control system

Motor rotation counterclockwise at collar speed

Collar rotation clockwise

Motor

Drive shaft at offset angle to collar

Drilling tendency

Figure 1-49. An electric motor is used by the PowerDrive Xceed* point-the-bit system to counter rotate a mandrel against the rotation of the collar. This keeps the mandrel, and thus the bit, oriented in the same direction while still rotating with the collar.

Figure 1-50. A geostationary offset angle between the bit and collar is used to create the steering tendency for the PowerDrive Xceed point-the-bit system.

1.6.1.2 Measurement-while-drilling technology

MWD tools have four major capabilities:

- real-time surveys for directional control—inclination, azimuth, toolface
- real-time power generation
- real-time mud-pulse data transmission telemetry system
- real-time drilling-related measurements—weight on bit, torque at bit, mud pressure.

Real-time wellbore position surveys and BHA orientation measurements

Knowledge of the well location in a formation and in 3D space is critical for all facets of well construction and formation evaluation. The inclination of a wellbore from vertical is determined by using a set of triaxial accelerometers to measure the components of the Earth's gravitational field. In conjunction with the inclination data, a set of triaxial magnetometers is used to measure the components of the Earth's magnetic field to determine the azimuth of the borehole with respect to magnetic north. Knowledge of the local horizontal angle between true and magnetic north (magnetic declination) enables conversion of the azimuth to a true north reference. In many cases a grid convergence correction is applied to account for local distortion resulting from the projection of the curved surface of the Earth onto a flat, 2D map. Azimuths corrected for this effect are said to be referenced to grid north.

During directional drilling, the orientation of the drilling system defines the direction in which the well deviates. Toolface is the angle between a reference, either gravity in a deviated well or north in a vertical well, and the direction in which the drilling assembly is oriented. Because the toolface is the direction in which the BHA tends to deviate the hole, it is used by the directional driller to ensure that the drilling assembly is oriented to give the desired direction to the well.

Real-time power generation

Batteries could be used to deliver power to the measurement electronics and telemetry system, but the duration of drilling runs would be limited to the life of the batteries. For this reason many MWD systems incorporate a downhole mud turbine and alternator as an electrical power generation system (Fig. 1-51). Whenever mud is being pumped through the drilling system, hydraulic power from the mud flow is converted into electrical power as the turbine rotates and drives the attached alternator. The generated electrical power is then available to the MWD subsystems and, where an intertool electrical connection is available that provides power and data connectivity along the BHA, power from the MWD turbo-alternator system can also be used by other tools in the BHA.

Figure 1-51. Hydraulic power from the mud flow is converted into electrical power through use of a turbine and alternator system. The stator (blue) deflects the mud flow (from left to right) on to the rotor (red) causing it to turn. The alternator (inside the yellow housing to the right) converts the rotation to electrical power.

Real-time mud-pulse telemetry

Real-time mud-pulse data telemetry techniques were originally developed to improve the efficiency of well-bore surveying while drilling, which had previously been acquired by the time-consuming process of running and retrieving single- or multishot mechanical surveying devices. The introduction of electronics and sensors capable of surviving the drilling environment, combined with the means to transmit the data to surface, has significantly reduced the time required to survey a well. The original MWD application, then, was the acquisition and transmission of survey measurements.

There are now several methods for transmitting data from downhole tools to the surface including electromagnetic propagation and wired drillpipe; however, the vast majority of real-time data transmission from downhole to surface is still performed by mud-pulse telemetry.

Mud-pulse telemetry involves encoding data in pressure pulses that propagate up through the mud inside the drillpipe. These pressure pulse sequences are detected at surface and decoded to recreate the numerical value of the data from the downhole tools. There are three main ways of creating a mud pressure pulse (Fig. 1-52). Positive pulse systems impede mud flow with a poppet valve, resulting in a temporary increase in pressure. Negative pulse systems use a bypass valve to bleed pressure off to the annulus, resulting in a temporary drop in pressure. A continuous carrier wave, or siren, system uses a rotating valve assembly that alternates between opened and closed positions, resulting in an oscillating pressure wave. Data can be encoded using the siren system by frequency, phase, or amplitude modulation.

Figure 1-52. The three major mud-pulse telemetry systems are the positive pulse, negative pulse, and siren, which generates a continuous carrier wave.

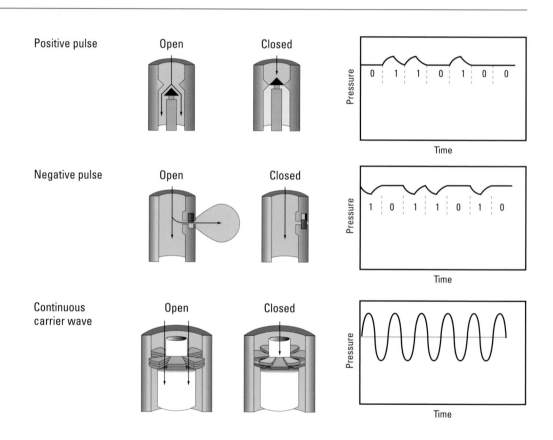

Real-time drilling-related measurements

Given a means to transmit data to surface it was not long before additional measurements such as GR and downhole drilling parameters were added to the downhole measurements available at the surface in real time. In some cases GR measurements provide the means to identify and correlate the formations being drilled to offset well data. Downhole measurements of weight on bit and torque at bit allow the driller to determine whether weight and torque are being smoothly transferred to the bit or whether friction between the drillstring and borehole wall is impeding the smooth transfer of mechanical power to the drilling interface.

The tools used to acquire formation evaluation measurements while drilling are generally referred to as logging-while-drilling tools to distinguish them from the drilling-oriented MWD tools. In general, LWD tools send selected formation evaluation data via an internal tool bus to the MWD tool for transmission to surface. All data transmitted in this way, along with the corresponding surface data acquired during drilling, is referred to as real-time (RT) data. Downhole tools also have memory in which all the measured data (as distinct from the limited selection of data sent in real time) is stored for retrieval when the tool returns to surface. Data extracted from the tool memory is referred to as recorded mode (RM) data.

Refer to Chapter 4, "Measurement-While-Drilling Fundamentals," for details on MWD technology.

1.6.1.3 Logging-while-drilling technology

When formation evaluation measurements were first migrated from wireline tools to drill collars, naturally the key triple-combo measurements of resistivity, density, and neutron response were the first to be made available in addition to the GR measurement, which was also available from the MWD tools. Owing to the nature of the drilling environment, in addition to the limitations imposed by having to fit detectors and electronics in a drill collar, some changes were made to the configuration of sensors. Because of these necessary changes, LWD tools and wireline tools measuring the same formation parameter generally have slightly differing raw responses. After appropriate environmental corrections are applied, any remaining discrepancies are generally the result of differences in the invasion profile and borehole condition between the early LWD log and the subsequent wireline log.

The evolution of LWD technology has seen significant improvements in the triple-combo services and the addition of numerous additional measurements, including imaging of multiple formation properties, magnetic resonance, sonic and seismic acquisition, elemental capture spectroscopy, and sigma measurements.

In addition to increasing measurement sophistication, LWD data has broadened in use from petrophysical evaluation of the formation to real-time measurements for evaluation of the wellbore location within the layers (well placement) and wellbore stability (geomechanics).

Measurement mode of conveyance question

Because wireline measurements are acquired after the hole is drilled, the cost of running the logs includes rig-time costs for the duration of the logging as well as the logging cost itself. The benefits of having formation data while drilling, in addition to the significant reduction in rig-time costs associated with running LWD rather than wireline logs, have resulted in LWD being used to acquire a significant proportion of openhole formation evaluation data.

The decision to run wireline logs or LWD logs is based on three major factors: rig cost, drilling risk, and well placement risk (Fig. 1-53). If any one of these is a high priority, then most likely LWD data will be acquired. If all are low in priority, then wireline data acquisition is most likely the preferred option.

Figure 1-53. The openhole formation evaluation market is served by LWD and wireline-conveyed measurements.

High rig cost	Low rig cost
or High drilling risk	*and* Low drilling risk
or High well placement risk	*and* Low well placement risk
Strong LWD value	Wireline preference
Wireline participation	Limited LWD value

In wells with high well placement risk, LWD measurements are required to deliver real-time information about the formation surrounding the borehole to identify the location of the well in the geological sequence. On the basis of this information the placement of the well can be optimized relative to the formation structure.

LWD measurement portfolio

To determine the location of a well within a geological sequence, measurements that enable the identification of markers must be available in real time. Markers used for steering are generally changes in a formation property, such as a high-GR layer just above the target reservoir or a change in formation resistivity that indicates where reservoir hydrocarbon saturation deteriorates. To detect these markers, LWD tools that measure the formation properties must be available. Over the years the LWD measurement portfolio has expanded significantly, with the following formation measurements are now available while drilling:

- Gamma ray
- Resistivity
  - Laterolog
  - Propagation
- Bulk density
- Photoelectric factor
- Neutron response
  - Thermal
  - Epithermal
- Magnetic resonance
  - Longitudinal relaxation time (T_1)
  - Transverse relaxation time (T_2)
  - Diffusivity of fluids in the pore space (D)

- Acoustic
  - Compressional
  - Shear
  - Stoneley
- Elemental spectroscopy
- Thermal capture cross section (sigma)
- Formation fluid analysis
  - Pressure
  - Optical fluid identification
  - Samples
- Seismic
  - Checkshot
  - Seismic profiling

- Remote boundary detection
  - Electromagnetic
- Images
  - Gamma ray
  - Laterolog resistivity
  - Bulk density
  - Photoelectric factor

Ongoing research and engineering efforts continue to enhance both the portfolio of measurements and the capabilities of existing measurements.

With such a wide selection of measurements available, a great deal of information about the formation surrounding the wellbore can be extracted in real time to assist in determining the location of the well in the geological sequence. In conjunction with the various well placement techniques, these measurements support enhanced wellbore positioning in addition to enhanced formation evaluation.

Refer to Chapter 5, "Logging-While-Drilling Fundamentals," for details on LWD.

1.6.2 Software and information technology

1.6.2.1 Real-time data transmission

Real-time transmission from downhole to the surface system

Data acquired by the downhole LWD tools is compressed, encoded, and transmitted to the surface, most commonly through a mud-pulse telemetry system. Owing to the limited bandwidth of current mud-pulse systems (typically 0.5 to 12 bits per second [bps]), the amount of data that can be transmitted to the surface in real time is limited. Improvements continue to be made in the telemetry rate (the number of bits that can be transmitted per second) and data compression (the amount of data transmitted per bit). Selection of what data is to be sent is still required because bandwidth capable of transmitting all the data all the time, as is the case with wireline tools, is unlikely to be widely available in the near future.

Real-time data is grouped into data points (d-points), each of which represents a particular measurement (for example, formation bulk density) or is part of a larger collection of data, such as part of an image that is spread across several d-points.

The d-points are grouped into frames, which define the data to be transmitted when the BHA is in a particular mode of operation (Fig. 1-54). For example, a frame designed for use when the BHA is sliding in a deviated well would contain GTF and nonazimuthal formation measurements because there is no point in wasting bandwidth by sending azimuthal data, which cannot be acquired when the BHA is not rotating. A frame designed for use when the BHA is rotating would, in contrast, most likely contain d-points for azimuthal and perhaps image data, but would not contain a toolface d-point because the toolface is not of interest when the BHA is rotating.

Real-time data is typically transmitted in four designated frames:

- MTF frame—used in nearly vertical wells when sliding

- GTF frame—used in deviated and horizontal wells when sliding

- rotary frame—used when the BHA is rotating

- utility frame—used to transmit data acquired while the mud pumps are off and hence the mud-pulse telemetry is not operational. This frame generally contains static surveys (acquired when the BHA is quiet) and hydrostatic mud pressure. It is increasingly used for applications such as acoustic (sonic and seismic) transit times and waveforms as well as formation pressure data acquired with the pumps turned off to minimize measurement interference.

Figure 1-54. The d-points containing measurement data are grouped into frames.

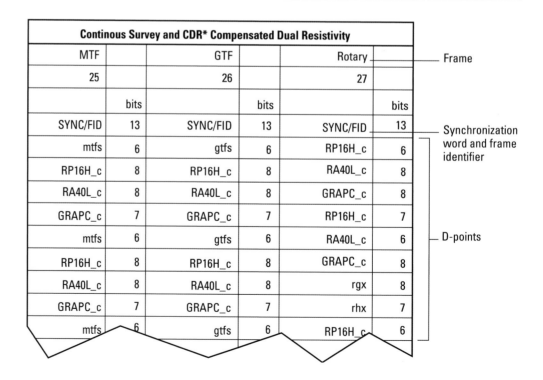

Continous Survey and CDR* Compensated Dual Resistivity						
MTF		GTF		Rotary		— Frame
25		26		27		
	bits		bits		bits	
SYNC/FID	13	SYNC/FID	13	SYNC/FID	13	Synchronization word and frame identifier
mtfs	6	gtfs	6	RP16H_c	6	
RP16H_c	8	RP16H_c	8	RA40L_c	8	
RA40L_c	8	RA40L_c	8	GRAPC_c	8	
GRAPC_c	7	GRAPC_c	7	RP16H_c	7	
mtfs	6	gtfs	6	RA40L_c	6	— D-points
RP16H_c	8	RP16H_c	8	GRAPC_c	8	
RA40L_c	8	RA40L_c	8	rgx	8	
GRAPC_c	7	GRAPC_c	7	rhx	7	
mtfs	6	gtfs	6	RP16H_c	6	

The selection of which frame to transmit is made by the downhole tool based on the well inclination (MTF or GTF frame), whether the tool is rotating (rotary frame), and whether the mud flow has just started again after a period of no flow (utility frame). After a few training bits are sent to allow the surface system to synchronize, the frame identifier is sent. Because both the surface and downhole systems have been programmed with the same frames, the subsequent stream of bits is divided into the corresponding d-points and decoded by the surface system.

Decoding involves the conversion of the binary bit stream to decimal followed by application of the reverse transform that was applied to the data downhole (Fig. 1-55). For example, a downhole density measurement of RHOB = 2.4 g/cm^3 could have 0.9 g/cm^3 subtracted and the remainder divided by 0.01 to give a decimal number of 150. The eight-bit binary equivalent, 10010110, is transmitted via the mud-pulse system to the surface, where it is converted back to the decimal number 150 and the reverse transform, RHOB_RT = 0.01X + 0.9 g/cm^3, is applied. The real-time RHOB_RT = 2.4 g/cm^3 measurement (the "_RT" suffix designates it as real-time data to distinguish it from the recorded-mode RHOB) is then available for visualization and interpretation.

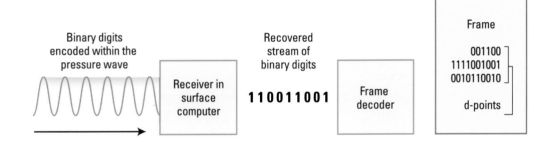

Figure 1-55. The bit stream encoded in the mud-pulse pressure wave is divided back into the d-points by the surface system, which is programmed with the same frame information as the downhole tool.

Real-time transmission from the surface system to the decision maker

If the real-time data user and decision maker are at the wellsite, then further transmission of the data from the acquisition system may not be immediately necessary. However, with the increasing use of remote operations support, it is likely that the data needs to be securely distributed to approved users. Data encryption and satellite transmission enable securely transferring the data from any rig to a satellite receiving station, from which it can be distributed over the Internet while still encrypted. Dedicated user accounts with password protection ensure that the data is available only to approved users.

In a few seconds, data acquired under the extreme conditions of downhole drilling can be made available to approved users anywhere in the world for subsequent interpretation and decision making.

1.6.2.2 Real-time information extraction

Once transmitted from the downhole tools to the surface and from the surface acquisition computer to the decision maker, the data must be presented in a manner that helps the decision maker extract the relevant information encoded in the data stream.

In the case of image data this is generally best achieved by 2D and 3D visualization of the data, color coded to represent the measured formation parameter. The addition of interactivity and dip-picking to the visualization environment allows the decision maker to extract quantitative information about the formation dip from the data stream. This capability facilitates well placement using the real-time dip determination technique.

The model, compare, and update method for well placement requires more sophisticated software support because the incoming real-time data must be displayed compared with modeled tool responses. The software must be able to create and modify a formation structural model populated with multiple formation properties and a planned well trajectory. In addition, the software must be able to simulate (forward model) the response of the LWD tools and stream in the real-time trajectory and logging data so that they can be compared with the simulated log response. As discussed in Section 1.5.1, "Model, compare, and update," discrepancies between the simulated and real data are an indication that the formation model does not accurately represent the subsurface formation and hence the model needs to be updated so that the simulated and actual data match. Once they match, the position of the wellbore in the formation can be assessed and appropriate well placement decisions taken and communicated to the directional driller.

Presentation of this data is generally of the form shown in Fig. 1-56. A curtain section of the formation along the planned well trajectory, color coded to show one of the formation properties of interest, is plotted in the lower panel with TVD as the vertical axis and THL as the horizontal axis. Formation structural information is captured in the scaled geometry of the curtain section layers and faults. The formation model can be color coded for any of the formation properties entered during the predrilling model construction.

Usually both the planned and real-time well trajectories are displayed so that any departure of the well from the planned well trajectory can be identified. Note that the curtain section is constructed during the predrilling preparation phase when only the planned well is available. Hence, the curtain section is typically a 2D vertical slice through a 3D formation model along the planned well trajectory. If the actual well trajectory departs significantly from the azimuth of the planned well trajectory, the curtain section may need to be reconstructed along the new well trajectory. More recent software allows log forward modeling and comparison in a 3D environment, removing the 2D constraint (Fig. 1-57).

Horizontal log tracks in the upper panel display both the real-time data and forward-modeled log responses so that discrepancies between them can be identified. Image data can also be displayed in the horizontal log tracks. If a discrepancy between the forward-modeled and real-time logs is identified, then interactive adjustments are made to the formation model. The simulated log responses are then recomputed for the edited formation model and compared again with the real-time data. This model, compare, and update cycle is repeated until a match is achieved.

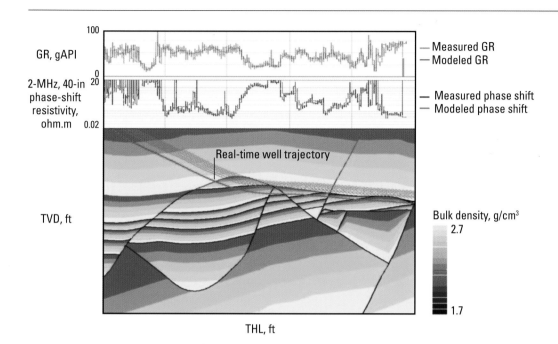

GR, gAPI

2-MHz, 40-in
phase-shift
resistivity,
ohm.m

— Measured GR
— Modeled GR

— Measured phase shift
— Modeled phase shift

Real-time well trajectory

TVD, ft

Bulk density, g/cm³
2.7

1.7

THL, ft

Figure 1-56. Forward-modeled and real-time results are compared in the horizontal log tracks. The position of the wellbore relative to layering in the property model is displayed in the lower panel.

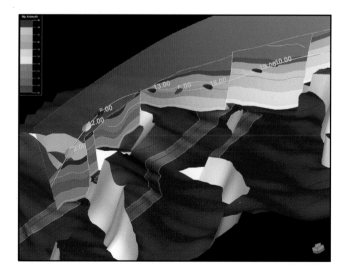

Figure 1-57. 3D representation of the formation and wellbore trajectory removes the 2D constraint that the curtain section is valid only if the planned and drilled wells are in the same vertical plane.

1.6.3 People and process

Technology and techniques are enablers, but well placement success is really built on the skills of the people who make it happen. Well placement is a highly collaborative exercise requiring good communication and teamwork.

1.6.3.1 The role of operating company geoscientists

Operating company geoscientists have the best understanding of the geology, depositional environment, and reservoir asset as a whole. This reservoir information is critical for correct placement of the well. Consider the example of well placement with a deep directional electromagnetic measurement that indicates an unexpected zone of higher resistivity above the borehole. The increased resistivity could result from lower porosity or higher oil saturation. The directional electromagnetic measurement cannot distinguish between these scenarios because it is sensitive only to the change in resistivity. The operating company geologist, with an understanding of the depositional environment, has the required knowledge to assess whether drilling into this zone is advisable.

Because operating company personnel use information that is often accessible only within their work environment, well placement is best performed in the operating company office. This not only ensures that the additional information is readily available, but also facilitates communication and helps the well placement team operate with mutual understanding of the objectives and operational constraints.

Well placement services provided by service companies have been misconstrued by some operating company personnel as an attempt to encroach on their responsibilities or even as a threat to their job security. This is certainly not the intention, because operating company geoscientists are an integral part of the well placement team whose understanding of the asset and corporate objectives is critical to ensure that well placement delivers value to the operating company.

1.6.3.2 The role of the well placement engineer

Whereas operating company personnel bring an understanding of the reservoir and corporate objectives, the well placement engineer brings detailed understanding of LWD technology and well placement techniques. The well placement engineer is a technical advisor on tool responses and interpretation to the operating company personnel. Before each operation, the well placement engineer must understand the operating company objectives and constraints in drilling the well, model the expected LWD measurement responses and ensure these are sufficient to achieve the well placement objectives, and advise the operating company personnel on the appropriate measurements and techniques. During operations, the well placement engineer is responsible for advising on steering decisions. Ultimately, any decision on adjustments to the well trajectory remains the responsibility of operating company personnel.

Consider the example discussed previously, in which the deep directional electromagnetic measurements indicate a layer of higher resistivity above the wellbore. The well placement engineer has the technical knowledge to perform quality control on the measurements and inversion results in consideration of the capabilities and limitations of the service. By presenting the two possible interpretations of the observed responses (higher resistivity above the well caused by either lower formation porosity or higher hydrocarbon saturation), the well placement engineer distills the technical information and suggests possible interpretations. Without requiring a detailed understanding of the service provider's technology, the required information is available for the operating company personnel to apply their skills. Within the operating company this ensures that the limited expertise on the asset is focused where it adds greatest value. Such collaboration is the key to successful well placement.

1.6.3.3 Team coordination

Well placement is a powerful technique because it brings together experts from different domains in both the operating company and service company organizations. Clear communication and coordination are vital for successful well placement.

An essential part of prejob preparation is the definition of roles and responsibilities for all involved. This includes operating company personnel, well placement engineers, directional drillers, and other wellsite personnel. Names, contact details (e-mail, office and mobile telephone numbers, fax numbers), responsibilities, and shift times should be documented and distributed to all involved prior to the start of the job.

The final decision maker must be clearly identified to all team members and contingencies planned should additional decision makers need to become involved, such as if the well needs to be abandoned or extended because of poor reservoir quality.

A record must be kept of all decisions and the time at which they were made. Secure Web communication chat facilities are an excellent communication tool because the author and time of each entry are logged. This is a valuable record to have if a review of operations is required.

Ultimately, team coordination is about ensuring that the information required to achieve the well objectives is available to the decision makers and the personnel responsible for executing the job.

Reservoir Geology Fundamentals

Geology *(from the Greek words* geo *[earth] and* logos *[study]) is the science of the Earth—its history, structure, composition, life forms, and the processes that continue to change it.*

Geology (from the Greek words *geo* [earth] and *logos* [study]) is the science of the Earth—its history, structure, composition, life forms, and the processes that continue to change it. Geology is a large and complex subject. The following is a very brief summary of the major points relevant to well placement.

In placing wells within a geological target it is important that the well placement engineer have a basic understanding of the geological processes that create reservoirs. This knowledge not only enables the well placement engineer to anticipate the most likely shape of the reservoir (for example, whether layers are straight or undulating) but also be able to discuss possible geological scenarios with the operating company geologist.

2.1 Rock classifications

There are three main classes of rock: igneous, metamorphic, and sedimentary. Of these, hydrocarbon reservoirs are typically formed in sedimentary rocks.

2.1.1 Igneous

Igneous rocks crystallize from molten Earth material, called magma, or lava when it is at the Earth's surface. Igneous rocks that crystallize slowly, typically below the surface of the Earth, are intrusive igneous rocks and have large, interlocking mineral crystals (coarse grained enough to see with the naked eye). Extrusive igneous rocks crystallize quickly at the Earth's surface and have small crystals (usually too small to see without magnification) or form a noncrystalline glass. Igneous rocks typically consist of the minerals quartz, mica, feldspar, amphibole, pyroxene, and olivine.

2.1.2 Metamorphic

Metamorphic rocks form from the alteration of preexisting rocks by changes in ambient temperature, pressure, volatile content, or all of these. Such changes can occur through the activity of fluids in the Earth and movement of igneous bodies or regional tectonic activity. The texture of metamorphic rocks can vary from almost homogeneous, or nonfoliated, to foliated rocks with a strong planar fabric, or foliation, produced by the alignment of minerals during recrystallization or by reorientation. Common metamorphic rocks are slate (metamorphosed shale) and marble (metamorphosed limestone). Graphite, chlorite, talc, mica, garnet, and staurolite are common metamorphic minerals.

2.1.3 Sedimentary

Sedimentary rocks are formed at the Earth's surface through the deposition of sediments derived from weathered rocks, biogenic activity, or precipitation from solution.

- Clastic sedimentary rocks such as conglomerates, sandstones, siltstones, and shales form as older rocks weather and erode and their particles accumulate and lithify, or harden, as they are compacted and cemented.

- Biogenic sedimentary rocks form as a result of shell building or other skeletal growth of organisms, such as coral reefs that accumulate and lithify to become carbonate, primarily limestone and dolomite.

- Precipitates form when solids concentrate as water is lost to evaporation. The evaporite minerals halite (salt) and gypsum can form vast thicknesses of rock as seawater evaporates.

Sedimentary rocks can include a wide variety of minerals, but quartz, feldspar, calcite, dolomite, and evaporite group and clay group minerals are most common because of their greater stability at the Earth's surface than that of many minerals in igneous and metamorphic rocks. Sedimentary rocks, unlike most igneous and metamorphic rocks, can contain fossils because they form at temperatures and pressures that do not obliterate fossil remnants.

The vast majority of the world's hydrocarbon reservoirs are in rocks of sedimentary origin.

2.2 The rock cycle

Just as the processes of evaporation, condensation, and precipitation move water through the water cycle, igneous, metamorphic, and sedimentary rock-forming processes move rock materials through the rock cycle (Fig. 2-1).

The dynamic nature of the Earth can change rocks from one type to another. Deep burial and melting create igneous rocks from preexisting rocks and sediments. Deformation and metamorphism produce metamorphic rocks from preexisting rocks and sediments. Uplift, erosion, and deposition can form sedimentary rocks from preexisting rocks and sediments.

Figure 2-1. The rock cycle changes rocks from one type to another as a result of the dynamic nature of the Earth.

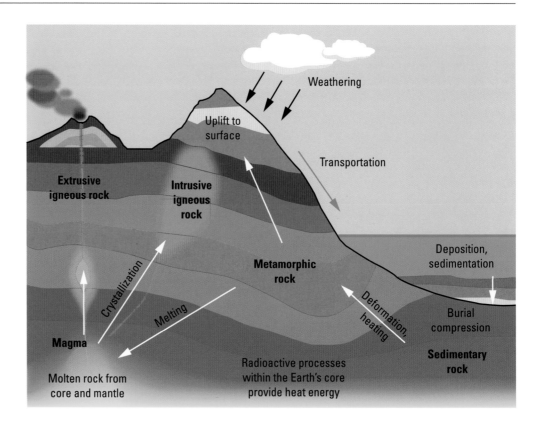

2.3 Depositional environments

Sediments are deposited under various conditions. Depositional environments describe the area in and physical conditions under which sediments are deposited, including the sediment source; depositional processes such as deposition by wind, water, or ice; and location and climate, such as desert, swamp, or river. The major depositional environments that create hydrocarbon reservoirs are shown in Fig. 2-2.

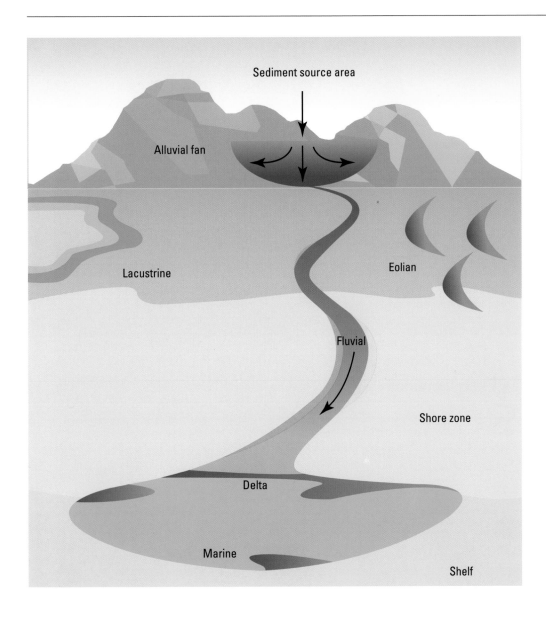

Figure 2-2. The depositional environments of sedimentary rocks include the sediment source and physical, chemical, and biological conditions and processes that affect the development of the rocks.

Alluvial deposition pertains to the subaerial (as opposed to submarine) environment, action, and products of a stream or river on its floodplain, usually consisting of detrital clastic sediments. It is distinct from subaqueous deposition, such as in lakes or oceans, and lower energy fluvial deposition. Sediments deposited in an alluvial environment can be subject to high depositional energy, such as fast-moving flood waters, and may be poorly sorted or chaotic.

Lacustrine deposition pertains to deposition in lakes or an area with lakes. Because deposition of sediment in lakes can occur slowly and in relatively calm conditions, organic-rich source rocks can form in lacustrine environments.

Eolian deposition pertains to the deposition of sediments by wind, such as sand dunes in a desert. Because fine-grained sediments such as clays are removed easily from wind-blown deposits, eolian sandstones are typically clean and well sorted.

Fluvial deposition pertains to deposition by a river or running water. Fluvial deposits tend to be well sorted, especially in comparison with alluvial deposits, because of the relatively steady transport provided by rivers.

Deltaic deposition pertains to an area of deposition or the deposit formed by a flowing sediment-laden current as it enters an open or standing body of water, such as a river spilling into a gulf. As a river enters a body of water, its velocity drops and its ability to carry sediment diminishes, leading to deposition. Sediments characteristically coarsen upward in a delta. The term originated with Herodotus in the 5th century BCE because the shape of deltas in map view can be similar to the Greek letter delta. The shapes of deltas are subsequently modified by rivers, tides, and waves. The three main classes of deltas are river dominated (for example, Mississippi River), wave dominated (Nile River), and tide dominated (Ganges River). Ancient deltas contain some of the largest and most productive petroleum systems.

Marine deposition pertains to deposition in seas or ocean waters, between the depth of low tide and the ocean bottom.

2.4 Plate tectonics

Plate tectonics is the unifying geologic theory that was developed to explain large-scale changes in the Earth that result from the interaction of the brittle plates of the lithosphere with one another and with the softer underlying asthenosphere (Fig. 2-3). The theory of plate tectonics initially stemmed from observations of the shapes of the continents, particularly South America and Africa, which fit together like pieces in a jigsaw puzzle and have similar rocks and fossils despite being separated by a modern ocean.

Plate tectonic theory explains such phenomena as earthquakes, volcanic or other igneous activity, mid-oceanic ridges and the relative youth of the oceanic crust, and the formation of sedimentary basins on the basis of their relationships to lithospheric plate boundaries. Convection of the mantle is postulated to be the driving mechanism for the movement of the lithospheric plates (Fig. 2-4). Global Positioning System measurements of the continents confirm the relative motions of plates. Age determinations of the oceanic crust confirm that it is much younger than continental crust because it has been recycled by the process of subduction, where one lithospheric plate moves beneath another, and regenerated at mid-oceanic ridges.

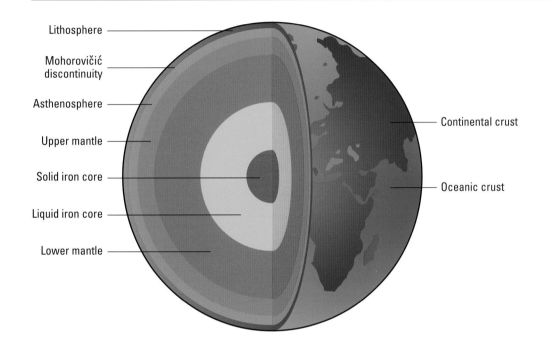

Lithosphere
Mohorovičić discontinuity
Asthenosphere
Upper mantle
Solid iron core
Liquid iron core
Lower mantle

Continental crust
Oceanic crust

Figure 2-3. The major compositional layers of the Earth are the core, mantle, and crust. The lithosphere includes all the crust and the uppermost part of the mantle. The asthenosphere forms much of the upper mantle. The Mohorovičić discontinuity (abbreviated to Moho) is the boundary between the crust and the mantle. These compositional and mechanical layers influence movement of the Earth's tectonic plates.

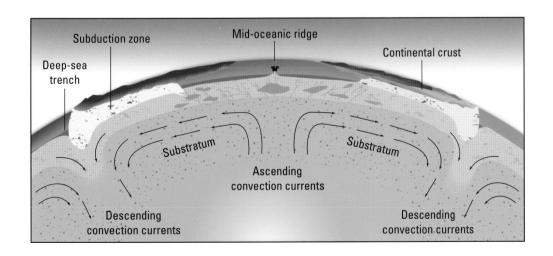

Subduction zone
Mid-oceanic ridge
Continental crust
Deep-sea trench
Substratum
Substratum
Ascending convection currents
Descending convection currents
Descending convection currents

Figure 2-4. Convection within the Earth has been proposed as a cause of the development of mid-oceanic ridges and subduction zones according to plate tectonic theory.

2.4.1 Divergent margins

Divergent margins occur where tectonic plates move apart. This splitting of the crust is known as rifting. Rifting is initiated when warm mantle material rises to beneath the crust, stretching and thinning the crust, which causes it to fracture and separate. The mantle material partially melts owing to the decrease in pressure and rises to fill the gap between the separating plates (Fig. 2-5).

At divergent margins,

- the region is initially uplifted as a result of thermal expansion (Fig. 2-5a; for example, Colorado plateau)
- normal faults and a rift valley develop as the plates begin to separate, with volcanism and earthquakes frequent (Fig. 2-5b, African Rift Valley)
- new oceans form as oceanic crust is produced (Fig. 2-5c, Red Sea)
- eventually a large ocean is created (Fig. 2-5d, Atlantic Ocean).

Figure 2-5. Divergent margins occur where tectonic plates move apart.

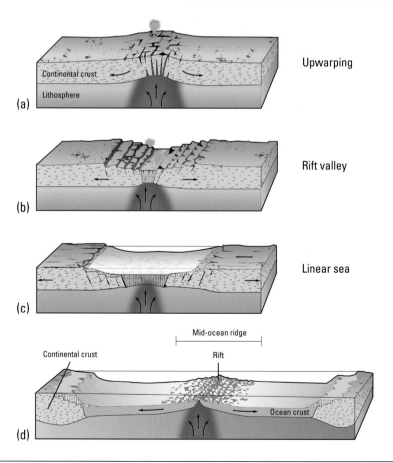

Extensive sedimentary deposits typically accumulate in the newly opened basin, with its large difference in relief between the uplifted horst blocks on either side and the downthrown graben, or valley, floor. As the rift expands, it may be periodically flooded by the sea, forming shallow-water clastic marine sediments interspersed with evaporites in hot, dry climates. Continued rifting eventually can open a new ocean. The two edges of the new continents become passive continental margins, no longer tectonically or volcanically active.

2.4.2 Convergent margins

Convergent margins occur where two or more tectonic plates or fragments of lithosphere move toward each other and form either a subduction zone or a continental collision (Fig. 2-6). In a subduction zone, one plate is subducted, moving beneath the other plate. Collisions between two continental plates can form large mountain ranges, such as the Himalayas.

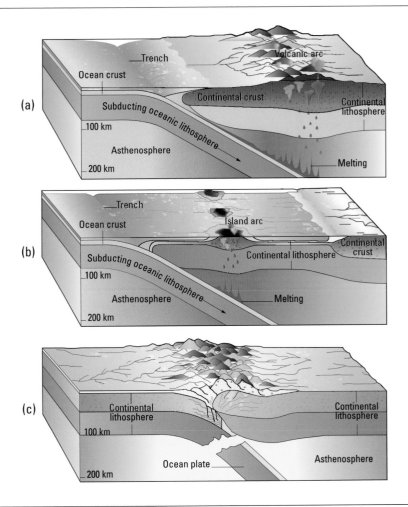

Figure 2-6. Convergent margins occur where tectonic plates move toward each other: (a) oceanic-continental plate convergence, (b) oceanic-oceanic plate convergence, and (c) continental-continental plate convergence.

2.4.3 Strike-slip margins

Strike-slip margins can occur at both divergent and convergent margins where tectonic plates slide and grind against each other along a slip-strike, or transform, fault. The relative motion of such plates is predominantly horizontal and usually triggers numerous strong, shallow earthquakes, such as along the San Andreas fault zone.

2.5 Faults

The dynamic nature of the Earth's crust results in the formation of faults and folds.

A fault is a break or planar surface in brittle (meaning having little or no inelastic deformation) rock across which there is observable displacement. Depending on the relative direction of displacement between the rocks, or fault blocks, on either side of the fault, its movement is described as normal, reverse, or strike-slip (Fig. 2-7).

Figure 2-7. Fault movement depends on the force-induced stresses on the rock.

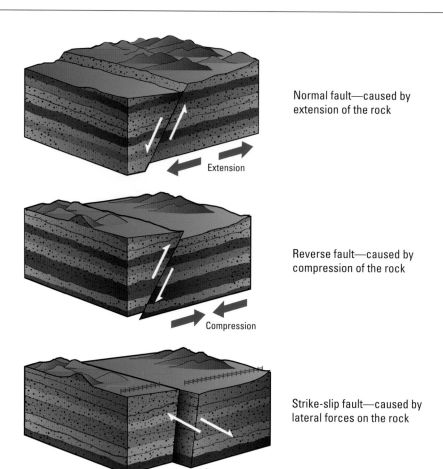

Normal fault—caused by extension of the rock

Extension

Reverse fault—caused by compression of the rock

Compression

Strike-slip fault—caused by lateral forces on the rock

Lateral

The fault block above the fault surface is called the hanging wall, whereas the block below the fault is the footwall. The throw of a fault is the vertical displacement of the layers caused by faulting (Fig. 2-8). The dip of a fault is the magnitude of the inclination of the fault plane from horizontal (angle α in Fig. 2-8).

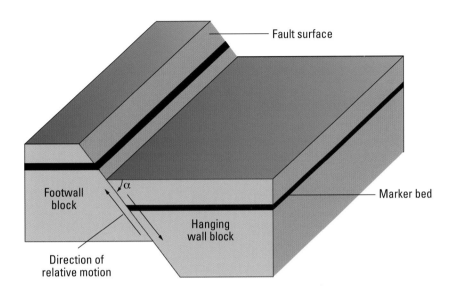

Figure 2-8. On a normal fault, the hanging wall has moved down relative to the footwall.

For a normal fault, the hanging wall moves down relative to the footwall along the dip of the fault surface, which is steep, from 45° to 90°. Multiple normal faults can produce horst and graben structures, which are a series of relatively high- and low-standing fault-bounded blocks that typically form in areas where the crust is rifting or being pulled apart by plate tectonic activity (Fig. 2-9). A growth fault is a type of normal fault along which movement occurs during sedimentation, which typically results in thicker strata on the downthrown hanging wall than on the footwall.

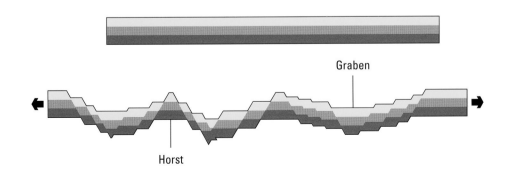

Figure 2-9. Horsts and grabens can form where the crust is being pulled apart.

A reverse fault forms when the hanging wall moves up relative to the footwall parallel to the dip of the fault surface. A thrust fault is a reverse fault in which the fault plane has a shallow dip, typically much less than 45°. In cases of considerable lateral movement, the fault is described as an overthrust fault. Thrust faults can occur in areas of compression of the Earth's crust.

Movement of normal and reverse faults can also be oblique, as opposed to purely parallel to the dip direction of the fault plane. The motion along a strike-slip fault, also known as a transcurrent or wrench fault, is parallel to the strike of the fault surface, and the fault blocks move sideways past each other. The fault surfaces of strike-slip faults are usually nearly vertical. A strike-slip fault in which the block across the fault moves to the right is termed a dextral strike-slip fault. If it moves left, the relative motion is called sinistral (Fig. 2-10). A transform fault is a type of strike-slip fault associated with convergent and divergent plate boundaries.

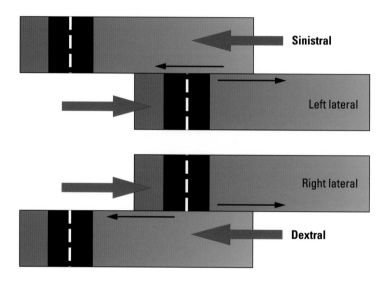

Figure 2-10. A road is offset by strike-slip faults, shown in plan view (looking down).

The presence of a fault can be detected by observing characteristics of the rocks such as changes in lithology from one fault block to the next, breaks and offsets between strata or seismic events, and changes in formation pressure in wells that penetrate both sides of a fault. Some fault surfaces contain relatively coarse rubble that can act as a conduit for migrating oil or gas, whereas the surfaces of other faults are smeared with impermeable clays or broken grains or have had impermeable minerals deposit in the fault plane, resulting in a sealing fault that can act as a seal.

Given the geological complexity of some faulted rocks, especially those that have undergone more than one episode of deformation, it can be difficult to distinguish between the various types of faults. Also, areas deformed more than once or that have undergone continual deformation can have fault surfaces that are rotated from their original orientations, so interpretation is not straightforward.

2.6 Folds

A fold is a wavelike geologic structure that forms when rocks deform by bending instead of breaking under compressional stress.

- Anticlines are arch-shaped folds in which rock layers are upwardly convex. The oldest rock layers form the core of the fold, and outward from the core the rocks are progressively younger.

- A syncline is the opposite of an anticline, with downwardly convex layers and young rocks in the core of the fold.

Folds typically occur in anticline-syncline pairs.

The hinge is the point of maximum curvature in a fold, with the limbs on both sides of the fold hinge. The hinge line connects the points of maximum curvature in a folded layer, and the axial surface connects the hinge lines of the layers in a fold (Fig. 2-11). The axial surface is called the axial plane for folds that are symmetrical and the hinge lines are coplanar.

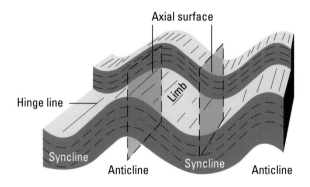

Figure 2-11. Folds are wavelike structures that typically have paired anticlines and synclines. Axial surfaces are imaginary surfaces that connect the hinge lines along the points of maximum curvature in a folded layer.

Concentric, or parallel, folding preserves the thickness of each layer as measured perpendicular to original bedding, whereas the layers of similar folds have the same wave shape but the thickness changes throughout each layer, with thicker hinges and thinner limbs.

A dome is a type of anticline that is circular or elliptical, with the rock layers dipping away in all directions. The upward migration of a mushroom- or plug-shaped salt diapir can form a salt dome. Similarly, a basin is a depression, or syncline, that is circular or elliptical rather than elongate (Fig. 2-12).

Figure 2-12. Domes and basins are the circular or elliptical forms of anticlines and synclines, respectively.

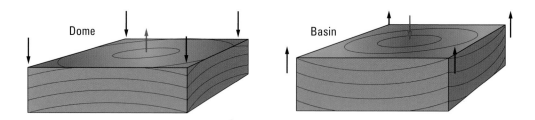

2.7 Petroleum systems

A petroleum system comprises the geologic components and processes necessary to generate and store hydrocarbons, including a mature source rock, migration pathway, reservoir rock, trap, and seal (Fig. 2-13). Appropriate relative timing of the formation of these elements and the processes of generation, migration, and accumulation are necessary for hydrocarbons to accumulate and be preserved. Exploration plays and prospects are typically developed in basins or regions in which a complete petroleum system has some likelihood of existing.

Source rock is rich in organic matter which, if heated sufficiently, generates oil or gas. Typical source rocks, usually shales or carbonates, contain about 1% organic matter and at least 0.5% total organic carbon (TOC), although a rich source rock might have as much as 10% organic matter. Preservation of organic matter without degradation is critical to creating a good source rock. Rocks of marine origin tend to be oil-prone, whereas terrestrial source rocks (such as coal) tend to be gas-prone.

Maturity refers to the state of a source rock with respect to its ability to generate oil or gas. As a source rock begins to mature, it generates gas. As an oil-prone source rock matures, the generation of heavy oils is succeeded by medium and light oils. Above a temperature of approximately 212 degF [100 degC], only dry gas is generated. The maturity of a source rock reflects the ambient pressure and temperature as well as the duration of conditions favorable for hydrocarbon generation.

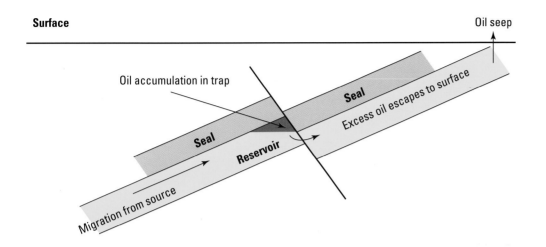

Surface

Oil seep

Oil accumulation in trap

Seal

Excess oil escapes to surface

Seal

Reservoir

Migration from source

Source kitchen area

Migration is the movement of hydrocarbons from their source into reservoir rocks. The movement of newly generated hydrocarbons out of their source rock is primary migration, also called expulsion. The further movement of the hydrocarbons into reservoir rock in a hydrocarbon trap or other area of accumulation is secondary migration. Migration typically occurs from a structurally low area to a higher area because of the relative buoyancy of hydrocarbons in comparison with the fluids in the surrounding rock. Migration can be local or can occur along distances of hundreds of kilometers in large sedimentary basins.

A reservoir is a subsurface body of rock with sufficient porosity and permeability to store and transmit fluids. Sedimentary rocks are the most common reservoir rocks because they have more porosity than most igneous and metamorphic rocks and form under temperature conditions at which hydrocarbons can be preserved.

Figure 2-13. A petroleum system must have a "kitchen," where mature source rock has reached appropriate conditions of pressure and temperature to generate hydrocarbons that migrate into a porous and permeable reservoir rock. The reservoir rock must have a seal to prevent the hydrocarbons from migrating through the reservoir rock. The reservoir rock and seal must be configured such that they form a trap in which the hydrocarbons accumulate. The timing of all these components is critical, as is the preservation of the reservoir seal and properties over time to ensure that the hydrocarbons are not lost.

A trap is a configuration of rocks suitable for containing hydrocarbons and sealed by a relatively impermeable formation through which hydrocarbons will not migrate. Traps are described as structural traps (in deformed strata such as folds and faults) or stratigraphic traps (in areas where rock types change, such as unconformities, pinchouts, and reefs) (Figs. 2-14 and 2-15).

Figure 2-14. Structural traps are formed by the deformation of strata such as faulting or folding.

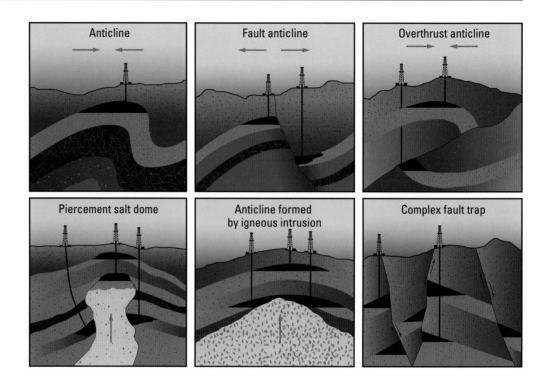

A seal is a relatively impermeable rock, commonly shale, anhydrite, or salt, which forms a barrier or cap above and around reservoir rock such that fluids cannot migrate beyond the reservoir. The permeability of a seal capable of retaining fluids through geologic time is typically ~10^{-6} to 10^{-8} D.

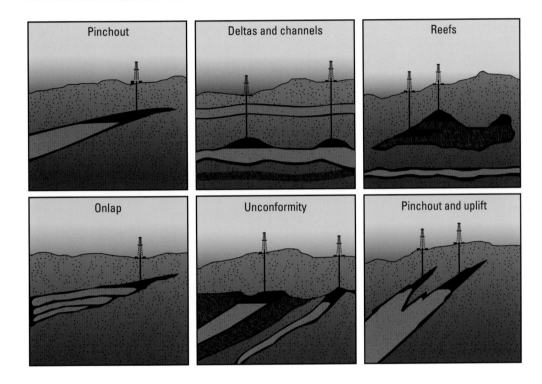

Figure 2-15. Stratigraphic traps are formed by erosional and sedimentary processes.

Timing refers to the sequence in which the components form and processes occur (Fig. 2-16). For example, if the source rock reaches maturity and generates oil but there is no trap formed over the reservoir rock, then the oil will eventually seep to surface and be lost to the atmosphere instead of accumulating.

Preservation refers to the stage of a petroleum system after hydrocarbons accumulate in a trap and are subject to degradation, remigration, tectonism, or other unfavorable or destructive processes.

As shown in Fig. 2-13, if the hydrocarbon volume that migrates into the reservoir exceeds the volume that can be held, then the excess hydrocarbons will spill from the trap. The spill point is the structurally lowest point in a hydrocarbon trap that can retain hydrocarbons. Once a trap has been filled to its spill point, further storage or retention of hydrocarbons cannot occur because of the lack of reservoir space within the trap. The hydrocarbons spill or leak out and continue to migrate until they are trapped elsewhere or seep to the surface.

Figure 2-16. The timing of the formation of the major elements of a petroleum system is critical in developing a hydrocarbon accumulation. The petroleum system events can be presented graphically to determine if the sequence may have resulted in hydrocarbon accumulations. This petroleum system is in the Maracaibo basin of Venezuela.[†] AAPG © 1994–1995, reprinted by permission of AAPG whose permission is required for further use.

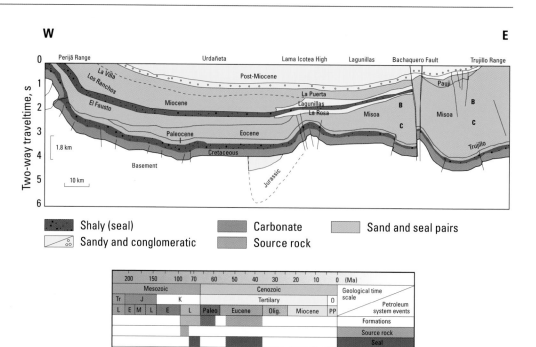

†Cross section after Parnaud, F., Gou, J., Pascual, J.-C., Capello, M.A., Truskowski, I., and Passlacqua, H.: "Stratigraphic Synthesis of Western Venezuela," *Petroleum Basins of South America*, A.J. Tankard, R. Suárez S., and H.J. Welsink (eds.), Tulsa, Oklahoma, USA, AAPG Memoir 62 (1995), p. 681. Stratigraphic column from Talukdar, S.C., and Marcano, F.: "Petroleum Systems of the Maracaibo Basin, Venezuela," *The Petroleum System—From Source to Trap* (1st ed.), L.B. Magoon and W.G. Dow (eds.), Tulsa, Oklahoma, USA, AAPG Memoir 60 (1994), p. 475.

2.8 Geologic terminology

The concepts of depth, thickness, and dip are fundamental for clear description of the subsurface. The various terminologies used are outlined in the following sections.

2.8.1 Depth

Depth is the distance from a reference point to a target point.

The depth reference is the point from which depth is measured, at which the depth is defined as zero. As shown in Fig. 2-17 the depth to a target horizon can be defined relative to mean sea level (MSL), ground level (GL) if on land, or the kelly bushing (KB) of the rig used to drill a well to the target horizon.

The depth reference for a well is typically the top of the KB or the level of the rig floor on the rig used to drill the well. The depth measured along the well trajectory from that point is the measured depth (MD) for the well. For Well A in Fig. 2-17, the MD to reach the target horizon is less than for Well B even though the target horizon is at the same true vertical depth (TVD). Even when the drilling rig has been removed, all subsequent measurements and operations in the well are still tied in to the original depth reference. However, for multi-well studies, the depths are usually shifted to the permanent datum, which is commonly MSL. Care must be taken in discussing the TVD of a target because drillers generally use the KB as their reference while geologists and geophysicists often have their maps referenced to MSL. Also, many horizontal wells are reentries, where a sidetrack is drilled out of an old well. In this case the rig that is drilling the current borehole may have a different rig floor or KB elevation compared with the rig that drilled the original borehole. This elevation offset must be taken into account when creating formation models and comparing MD and TVD logs.

Information regarding reference elevations such as GL above MSL and KB and drill floor above GL or MSL is recorded on the log heading.

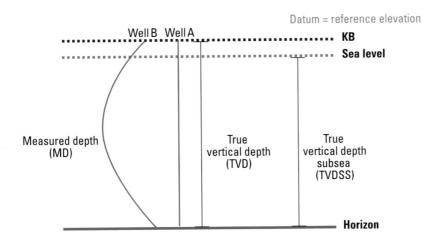

Figure 2-17. Depth is the distance from a reference to a target point. Both the reference and path to the target must be defined.

2.8.2 Thickness

Four terms are commonly used in describing the thickness of a layer:

- true stratigraphic thickness (TST) or true bed thickness (TBT)
- true vertical thickness (TVT)
- measured depth thickness (MDT)
- true vertical depth thickness (TVDT).

True stratigraphic thickness is the thickness of a rock layer measured perpendicular to the layer. Although this is the most intuitive definition of thickness, determining the TST in a well requires corrections for the dip and azimuth of both the layer and the well that intersects it.

The true vertical thickness is the thickness measured vertically at a point (Fig. 2-18). TVT is the thickness that is measured in a vertical well through the layer.

Figure 2-18. True stratigraphic thickness is the thickness of a rock layer measured perpendicular to the layer, whereas true vertical thickness is the thickness measured vertically at a point.

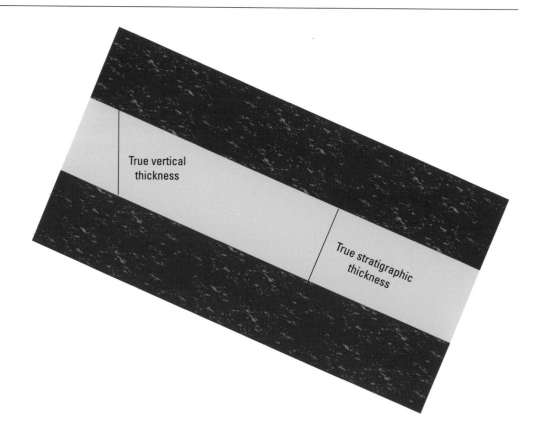

True vertical thickness

True stratigraphic thickness

The TST values in an area can be plotted and contours drawn to create an isopach map (see Section 2.9, "Contour maps"). The TVT values in an area can be plotted and contours drawn to create an isochore map.

The measured depth thickness of a layer is the thickness of the layer measured along the trajectory of the well that intersects that layer. Although the TST or TBT of a layer can remain constant, the MDT of the layer intersected by a wellbore varies as a function of the dip and azimuth of the layer as well as the dip and azimuth of the wellbore that intersects it.

Figure 2-19 shows a dipping layer intersected by three wells. Ignoring azimuthal effects in or out of the page, Well A, which is drilled perpendicular to the layer dip, has an MDT equal to the TST. Well B is vertical, so in this well the MDT of the layer equals the TVT. To convert the TVT to the TST, the dip of the layer must be taken into account.

Well C, which is drilled at a low incidence angle between the well and layer, has the longest exposure to the layer and hence the largest value of MDT.

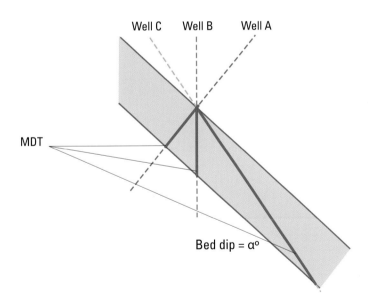

Figure 2-19. The measured depth thickness (solid red lines) of a layer increases with decreasing incidence angle between the wellbore and the layer in which the MDT is determined.

As shown in Fig. 2-20 for a well drilled perpendicular to the layer, the MDT and TST are equal:

$$TST = TVT(\cos \alpha),\qquad(2\text{-}1)$$

where α = true formation dip.

Figure 2-20. For a well drilled perpendicular to the layer, the MDT and TST are equal. A vertical well through a layer has an MDT that equals the layer TVT at that point.

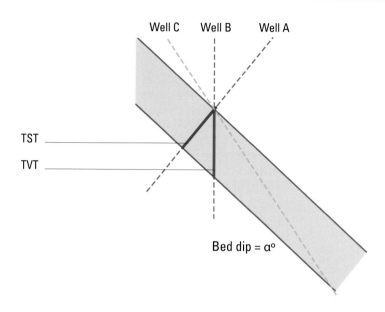

Converting an MD log to a TVD index does not result in the layers appearing in TVT. Consider the case shown in Fig. 2-21, where the true vertical depth thickness of the layer in Wells A and C is shown in black. For Well B, which is vertical, the TVT and TVDT values of thickness are the same.

Because Well C is drilling downdip along the layer, its MDT is greater and the vertical projection of the MDT is also greater than that in Wells A and B. A TVD log from Well C shows this layer as considerably thicker than a TVD log from either of the other wells. To convert from TVDT to TVT, the incidence angle and azimuth angle between the wellbore and layer must be taken into account:

$$TVT = MDT[\cos \beta - (\tan \alpha)(\sin \beta)(\cos \gamma)],\qquad(2\text{-}2)$$

where
β = well deviation
γ = acute angle between well azimuth and dip azimuth.

Accounting for change in TVDT is particularly important when correlating between wells. If one well has been drilled updip and another drilled downdip, then when the MD log is converted to TVD for correlation the layer thicknesses will be different (Fig. 2-22).

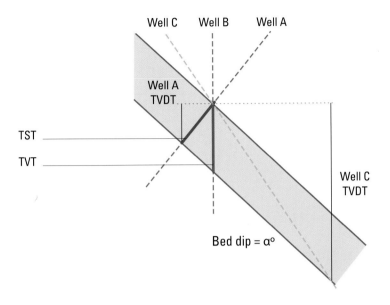

Figure 2-21. True vertical depth thickness is the vertical component of the MDT. It is not the same as the TVT.

From Fig. 2-22 it can be seen that for a well drilling updip, TVDT < TVT < TST. For a well drilling downdip in the same formation, TVDT > TVT > TST. The TST remains the same in both wells, but when projected to a TVD depth scale, the layer thicknesses appear different.

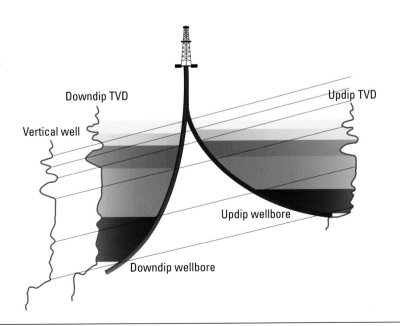

Figure 2-22. Layer thicknesses appear different on a TVD log from a well drilled downdip compared with a TVD log from a well drilled updip. Well-to-well correlations must account for this effect.

2.8.3 Formation dip

As outlined in Section 1.5.2.6, "Strike, dip, and azimuth," formation dip is the angle between a planar feature, such as a sedimentary bed or a fault, and a horizontal plane. True dip is the maximum angle a plane makes with a horizontal plane, with the angle measured in a direction perpendicular to the strike of the plane (Fig. 2-23).

Projected dip, also known as apparent dip, is the angle measured in any direction other than perpendicular to the strike of the plane. Given an apparent dip and the strike, or two apparent dips, the true dip can be computed (Fig. 2-24).

$$\tan \delta = (\tan \alpha)(\cos \beta), \qquad\qquad (2\text{-}3)$$

where

α = true dip of the layer

β = angle between the azimuth of the dipping plane and the azimuth of the apparent dip

δ = apparent dip of the layer.

Equation 2-3 can be proved from Fig. 2-24:

$AB = CD$

$\tan \delta = CD/OC = AB/OC$

$AB = OA\tan \alpha$

$OC = OA/\cos \beta.$

Substituting for AB and OC,

$\tan \delta = (OA\tan \alpha)/(OA/\cos \beta).$

Therefore, $\tan \delta = (\tan \alpha)(\cos \beta).$

For example, a well drilled at 45° from the azimuth of maximum (true) dip shows an apparent dip along the well of 10°. The true dip α is derived as

$\tan \alpha = \tan \delta/\cos \beta$

$\qquad = \tan 10°/ \cos 45°$

$\qquad = 0.249.$

Therefore, $\alpha = 14°.$

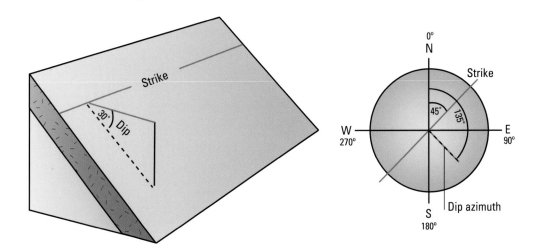

Figure 2-23. Strike is the azimuth of the intersection of a plane, such as a dipping bed, with a horizontal surface. Dip is the magnitude of the inclination of a plane from horizontal. True, or maximum, dip is measured perpendicular to strike.

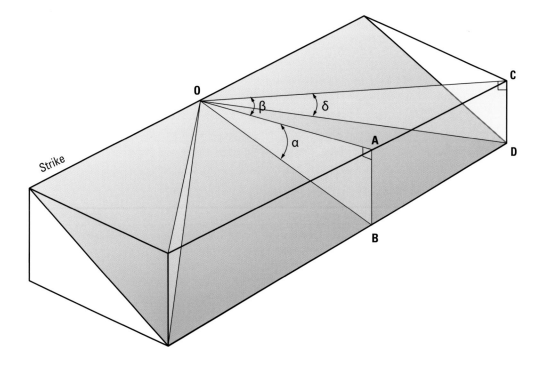

Figure 2-24. The true dip α is measured perpendicular to the strike; apparent dip δ is measured at any other angle.

2.9 Contour maps

Contour maps display lines that connect points of equal value to separate points of higher value from points of lower value. Common types of contour maps (Table 2-1) include

- topographic contour maps—show the elevation of the Earth's surface

- structure contour maps—show the elevation or depth of a formation

- gross or net reservoir or pay maps—show variations in the thickness of a stratigraphic unit, also called isopachs.

Table 2-1. Contour Maps Display Different Types of Data

Contour Map	Data
Isobar	Pressure
Isochore	True vertical thickness of formation or pay
Isopach	True stratigraphic thickness of formation or pay
Isotherm	Temperature
Structure	Elevation of formation
Topographic	Elevation at the surface of the Earth

The contour map most commonly used in well placement is the structure map. A structure map is a type of subsurface map with contours representing the elevation of a particular formation, reservoir, or geologic marker in space, such that folds, faults, and other geologic structures are clearly displayed (Fig. 2-25). Its appearance is similar to that of a topographic map, but a topographic map displays elevations of the Earth's surface and a structure map displays the elevation of a particular rock layer, generally beneath the surface. Structural elevation information is generally derived from the interpretation of seismic reflectors and formation tops identified in well logs.

The contours are labeled with negative numbers to indicate the true vertical depth of the formation top below mean sea level. The smooth contours shown in Fig. 2-25 indicate that there is no faulting identified in this double-domed structure.

Displacement along a fault results in discontinuous contour lines. Figure 2-26 shows the structure map and 3D representation of a simple dome structure with a planar fault that has not yet moved. The fault is shown as a curve across the contour lines. This is because the planar fault is dipping, so it intersects the contour lines along an arc. A vertical fault would appear as a straight line on the structure map.

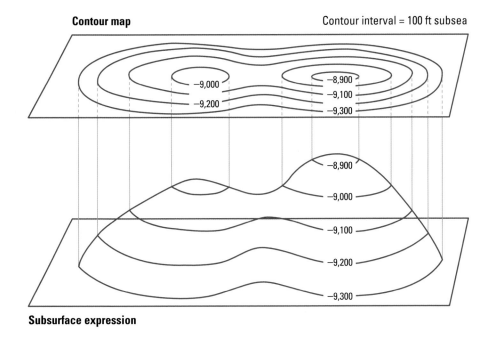

Contour map
Contour interval = 100 ft subsea

−9,000
−8,900
−9,100
−9,200
−9,300

−8,900
−9,000
−9,100
−9,200
−9,300

Subsurface expression

Figure 2-25. A structure map represents the subsurface elevation of a geologic surface.

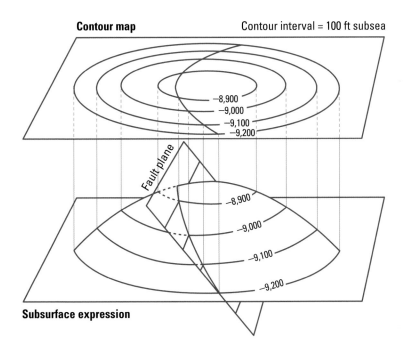

Contour map
Contour interval = 100 ft subsea

−8,900
−9,000
−9,100
−9,200

Fault plane

−8,900
−9,000
−9,100
−9,200

Subsurface expression

Figure 2-26. The structure map and 3D representation of a simple dome structure show a fault plane, but without any movement along the fault plane.

Figure 2-27 shows the same faulted dome structure after the structure has moved along the fault plane. In the 3D representation the contour lines remain at the same level, but the right side of the structure has dropped. The top contour line on the right structure is now −9,000 ft, rather than the −8,900 ft it was before movement along the fault. However, the −9,000-ft contour can be followed around the two sides of the dome and across the face of the fault.

Structure maps often indicate where wells intersect the formation and the status of the wells (Figs. 2-28 and 2-29).

If a horizontal well is drilled parallel to a contour line, then the apparent dip of the formation in the well is zero because the well is drilled along strike. Drilling perpendicular to the contour lines results in the well intersecting the true dip of the formation. Any other well azimuth results in the well intersecting an apparent dip that varies with the angle between the well azimuth and the azimuth of the formation dip (Fig. 2-24).

Figure 2-27. The same faulted dome structure shown in the previous figure has had displacement along the fault plane.

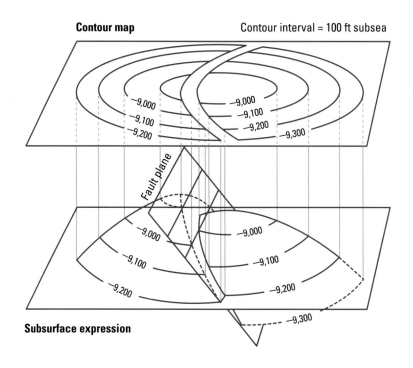

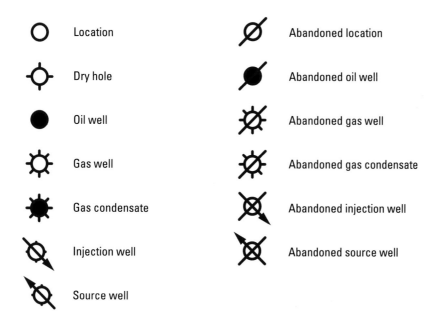

○	Location	∅	Abandoned location
ⴲ	Dry hole	⬤̸	Abandoned oil well
⬤	Oil well	☼̸	Abandoned gas well
☼	Gas well	☀̸	Abandoned gas condensate
✹	Gas condensate	⊗→	Abandoned injection well
ⴲ→	Injection well	⊗↙	Abandoned source well
ⴲ↙	Source well		

Figure 2-28. Well status is indicated by symbols on field structure maps.

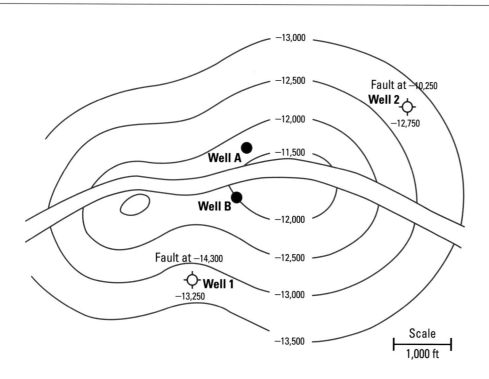

Figure 2-29. A typical structure map of a small oilfield has contour lines at a 500-ft interval. There is a fault across the middle of the structure, two producing oil wells on the crest of the structure, and two dry holes on the flanks of the structure.

Directional Drilling Fundamentals

Directional drilling *is the intentional deviation of a wellbore from the path it would naturally take.*

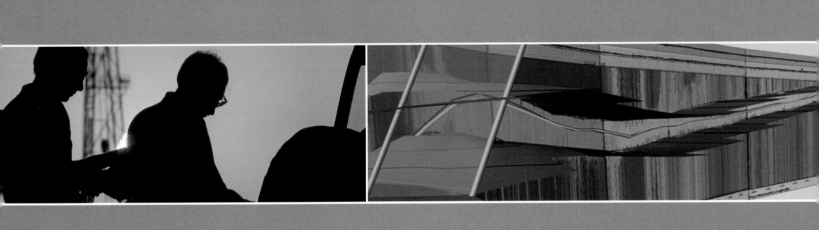

3.1 Definition

Directional drilling is the intentional deviation of a wellbore from the path it would naturally take.

In addition to being drilled directionally, the wellbore must

- be drilled as safely as possible

- be placed in the required target

- facilitate the planned logging program

- allow smooth running of casing and completion hardware

- not result in excessive casing wear from subsequent operations

- be accessible for future well intervention

- be drilled at the lowest possible cost.

Directional drilling is accomplished through the use of whipstocks, BHA configurations, instruments to measure the path of the wellbore in 3D space, data links to communicate measurements taken downhole to the surface, mud motors, and special BHA components and drill bits.

The directional driller also exploits drilling parameters such as weight on bit and rotary speed to deflect the bit away from the axis of the existing wellbore. In some cases, such as drilling steeply dipping formations or unpredictable deviation in conventional drilling operations, directional drilling techniques may be employed to ensure that the hole is drilled vertically.

Although there are many directional drilling techniques, the general concept is simple: orient the bit in the direction that one wants to drill. The most common way is through the use of a bend near the bit in a down-hole steerable mud motor. The bend points the bit in a direction different from the axis of the wellbore when the entire drillstring is not rotating. By pumping mud through the mud motor, the bit turns while the drillstring does not rotate, allowing the bit to drill in the direction it points. When a particular wellbore direction is achieved, that direction may be maintained by rotating the entire drillstring (including the bent section) so that the bit does not drill in a single direction off the wellbore axis, but instead sweeps around and its net direction coincides with the existing wellbore. Rotary steerable systems (RSSs) enable steering while rotating, usually with higher rates of penetration and ultimately smoother boreholes.

3.2 Reasons for drilling directionally

There are numerous reasons for drilling directionally, as shown in Figs. 3-1 through 3-8.

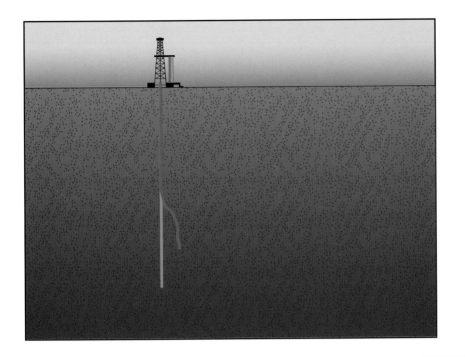

Figure 3-1. Sidetracking avoids undrillable material left in the hole (such as a "fish") so that drilling can continue.

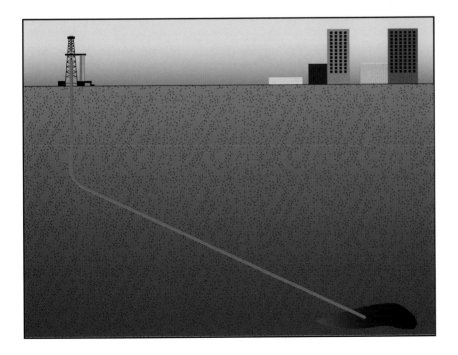

Figure 3-2. Accessing locations where a vertical well is undesirable includes directional drilling under populated or environmentally sensitive areas, or from land to locations below water to avoid having to install offshore facilities.

Figure 3-3. Cluster drilling is employed to keep the surface footprint to a minimum. Drilling multiple wells from an offshore platform is one example. Increasingly, land wells are being drilled from centralized clusters to minimize the environmental impact of wellheads from vertical wells scattered over a field.

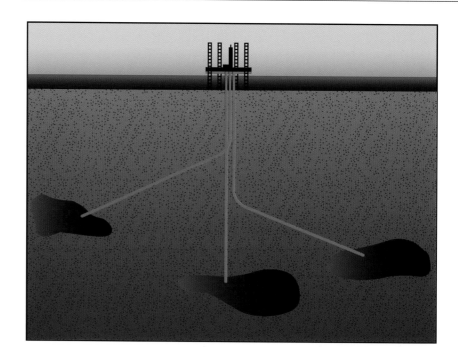

Figure 3-4. Salt dome drilling is used where an unstable overhanging salt formation prevents accessing the hydrocarbons with a vertical well.

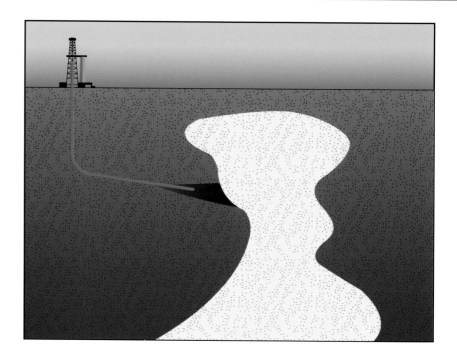

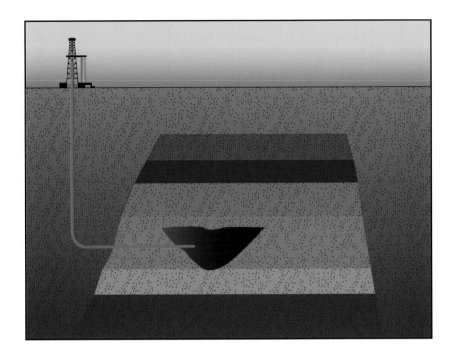

Figure 3-5. Fault control drilling minimizes well contact with a fault by drilling perpendicular to the fault plane.

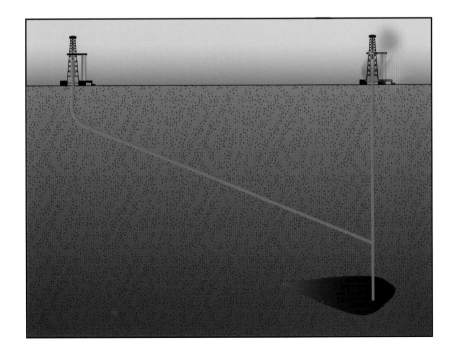

Figure 3-6. Relief well drilling relieves the pressure on a well out of control at the surface by drilling an additional well near or into the target well and producing it to reduce the pressure and hence flow into the target well.

Figure 3-7. Horizontal well drilling has the objective of maximizing the length of the well in the reservoir.

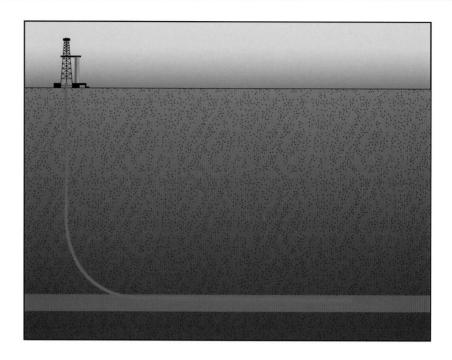

Figure 3-8. Multilateral well drilling achieves even greater reservoir exposure than a horizontal well by drilling multiple lateral wells from one main wellbore. Multiple reservoirs can be accessed or greater exposure provided to a single reservoir.

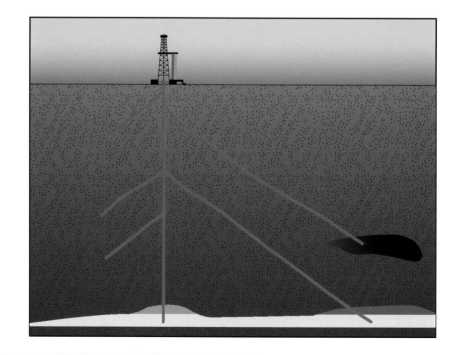

3.3 Well deflection methods

Since the introduction of planned directional drilling, a number of techniques have been developed including jetting, whipstocks, and rotary steerable assemblies. The two most commonly used techniques for well placement are the steerable motor and RSS.

3.3.1 Jetting

For jetting, the drill bit is usually equipped with small-diameter tungsten carbide nozzles, which produce a high-velocity drilling fluid stream exiting the bit. The jetting action cleans the bit and agitates the formation cuttings into suspension in the mud for transport to the surface and out of the hole. Though not as common as in the past, a bit may be fitted with asymmetric nozzles, one large and two or more small nozzles. If drillstring rotation is prevented during the jetting operation, the different nozzle sizes cause greater erosion on the side where the large nozzle is, enabling intentional deviation of the well (Fig. 3-9).

3.3.2 Openhole whipstock

An openhole whipstock is an inclined wedge placed in a wellbore to force the drill bit to start drilling in a direction away from the established wellbore axis (Fig. 3-10). The whipstock must have hard steel surfaces so that the bit preferentially drills through the rock rather than the whipstock itself. Whipstocks can be oriented in a particular direction if needed or placed into a wellbore blind, with no regard to the direction they face. Most whipstocks are set on the bottom of the hole or on top of a high-strength cement plug.

Figure 3-9. Jetting can be used to erode one side of the hole, resulting in a tendency for the BHA to deflect in the direction of the "pocket."

Figure 3-10. The wedge-shaped whipstock is used to direct the bit away from the axis of the established borehole to drill a deviated well.

Figure 3-9

Figure 3-10

3.3.3 Rotary steerable assembly

A stabilized BHA can be designed to build, hold, or drop angle depending on the location of the three points of contact nearest to the bit (touch points) between the BHA and formation (Fig. 3-11).

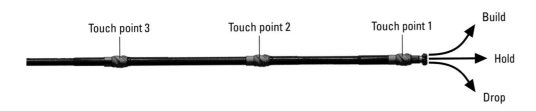

Figure 3-11. The positioning and diameter of touch points on a BHA defines whether the assembly tends to increase (build), maintain (hold), or decrease (drop) well inclination.

In a deviated hole gravitational and buckling forces bend any unsupported section of the BHA. If two stabilizers are widely spaced with one of them relatively close to the bit, then the BHA bends between them, resulting in the bit deflecting upward and the well inclination increasing. This is called a build assembly (top panel of Fig. 3-12).

If stabilizers are positioned a significant distance from the bit, the length of BHA between the bottommost stabilizer and the bit bends slightly under gravity, resulting in a tendency for the bit to point down and decrease the well inclination. This is called a pendulum or drop assembly (bottom panel of Fig. 3-12).

If stabilizers are distributed relatively evenly along the length of the BHA, then there is no tendency to either build or drop. This assembly tends to hold the well inclination and is called a packed or hold assembly (middle panel of Fig. 3-12).

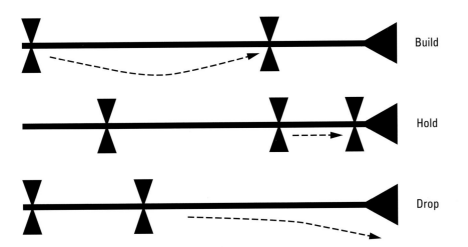

Figure 3-12. Unsupported sections of a BHA bend under the influence of gravity and buckling loads caused by high weight on bit. The bending can result in a deflection of the bit (shown by the dashed lines) and hence the inclination at which the well is drilled.

The degree to which an assembly builds, holds, or drops is called the build rate, which is the rate of change in borehole inclination given in degrees per 100 ft [30 m]. The degree to which an assembly turns to the right or left is called the turn rate and is also expressed in degrees per 100 ft [30 m]. The dogleg severity (DLS) of a well is the vector sum of the build and turn rates. DLS accounts for the change in both the inclination and azimuth expressed in degrees per 100 ft [30 m].

In addition to the positioning and diameter of touch points, a number of other factors influence the dogleg tendency of a BHA.

Larger diameter collars with thicker walls are stiffer and hence do not bend as easily as smaller diameter or thinner walled collars (Fig. 3-13). Because of the greater flexibility of a 4¾-in assembly compared with that of a 6¾-in or larger assembly, 4¾-in assemblies can deliver greater doglegs, but they can also be more challenging with respect to maintaining directional control.

Figure 3-13. Increased BHA element flexibility increases dogleg tendency.

Increasing the weight on bit enhances collar deflection, thereby increasing the build or drop tendency of the BHA (Fig. 3-14). The touch-point locations and element flexibility are fixed once the BHA is run in the hole. Changing the weight on bit is one of the few parameters that the driller can vary to modify the dogleg that an assembly produces in a particular formation.

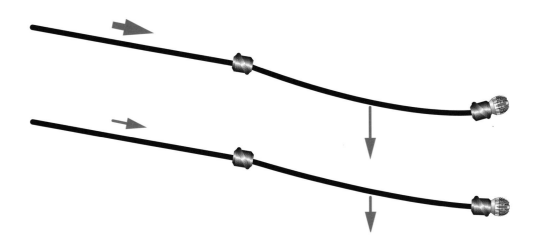

Figure 3-14. Increased weight on bit increases the dogleg tendency.

3.3.4 Steerable motors

Steerable motors consist of a positive displacement motor (PDM) with a surface-adjustable bent housing that enables orienting the bit in the desired drilling direction. The system comprises six key elements (Fig. 3-15):

- top sub—acts as a crossover from the power section to the drillstring

- power section—has a rotor and stator, which convert mud flow into bit rotation

- transmission section—transmits the rotation from the power section to the drive shaft

- surface-adjustable bent housing—enables setting the bend in the motor housing

- bearing section—supports all the axial and radial loads on the drive shaft on bearings

- drive shaft—is driven by the power section through the transmission section and has the drill bit screwed into a bit box at the bottom of the motor.

Figure 3-15. Steerable motors consist of six key elements, as shown on the cross section. The longest element of the steerable motor is the power section.

Top sub

Power section

Transmission assembly

Surface-adjustable bent housing

Bearing section

Drive shaft

The power section, also known as the PDM, converts hydraulic power from the mud circulation into mechanical power in the form of bit rotation (Fig. 3-16).

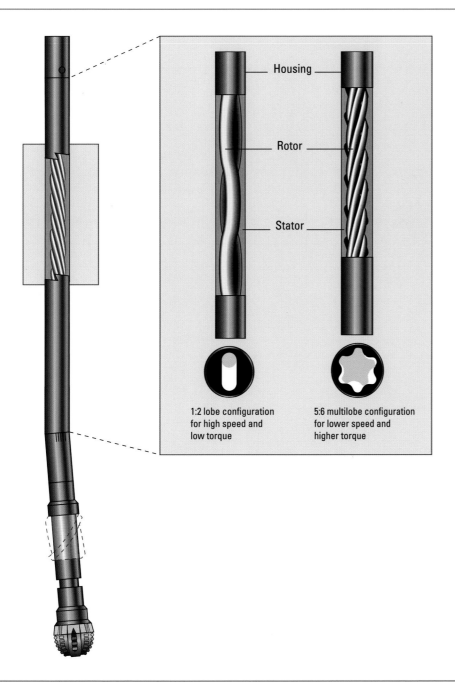

Figure 3-16. A PDM converts hydraulic power from the mud flow into mechanical power in the form of bit rotation.

Housing

Rotor

Stator

1:2 lobe configuration for high speed and low torque

5:6 multilobe configuration for lower speed and higher torque

The power conversion is achieved through a progressing cavity design in which the movement of the mud pushes on a rotor with one less lobe than the number of cavities in the stator in which it is housed (Fig. 3-17).

The movement of mud from one cavity to the next turns the rotor and thus turns the bit coupled to the rotor through the connecting rod and drive shaft. The key feature of a PDM is that the bit rotates when there is mud circulation, even if the drillstring is stationary.

Increasing the number of lobes and cavities increases the torque available but decreases the speed of bit rotation. Consider a unit volume of mud flowing through a motor. If the motor has a 1:2 configuration, the rotor progresses $360°/2 = 180°$. If the motor has a 7:8 configuration, then the rotor progresses $360°/8 = 45°$ for the same volume of mud. For a given mud flow rate, the revolutions per minute (rpm) of a motor is mainly controlled by the rotor-stator configuration.

Similar to a gearbox in a car, higher gearing ratios give lower rpm and higher torque. As shown plotted in Fig. 3-18, torque increases and the rpm decrease with an increase in the number of lobes and cavities in the motor configuration.

Figure 3-17. The cross section of a motor looking down the length of the motor (left) shows the fit of the rotor that has one less lobe than the corresponding stator has cavities, leaving a gap for mud circulation to push on the spiraled rotor and create rotation, as shown in the cut-away section of a motor (right).

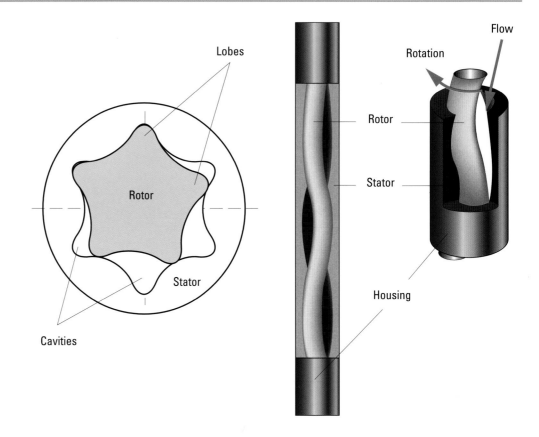

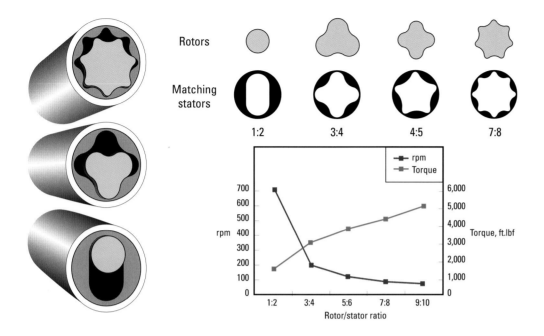

Rotors

Matching stators

1:2   3:4   4:5   7:8

Figure 3-18. Increasing the number of lobes and cavities increases the torque available with a corresponding decrease in the bit rpm.

The rotor has a highly polished surface that creates a seal with the rubberized elastomer insert forming the internal profile of the stator. The elastomer is housed in the stator tube, or "can." The dimensions of the rotor and stator must be carefully matched to ensure a good seal while preventing excessive interference between the rotor and stator (Fig. 3-19).

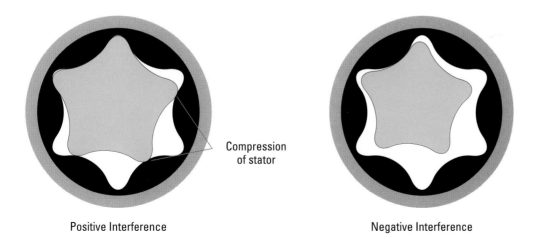

Positive Interference   Negative Interference

Compression of stator

Figure 3-19. Correct fit of the rotor to the stator is vital to motor performance. Negative interference results in loss of power; excessive positive interference results in rapid wear and heat generation.

Pressure differential across two adjacent cavities forces the rotor to turn, opening adjacent cavities and allowing the fluid to progress down the length of the stator (Fig. 3-20).

Figure 3-20. A cross section along the length of the motor (upper panel) shows how the spiraled rotor seals on the elastomer inside the stator tube so that the mud flow must turn the rotor to gain access to the next cavity. The 3D wireframe (lower panel) shows how the mud-filled cavity (red) moves along the motor as the rotor turns inside the stator.

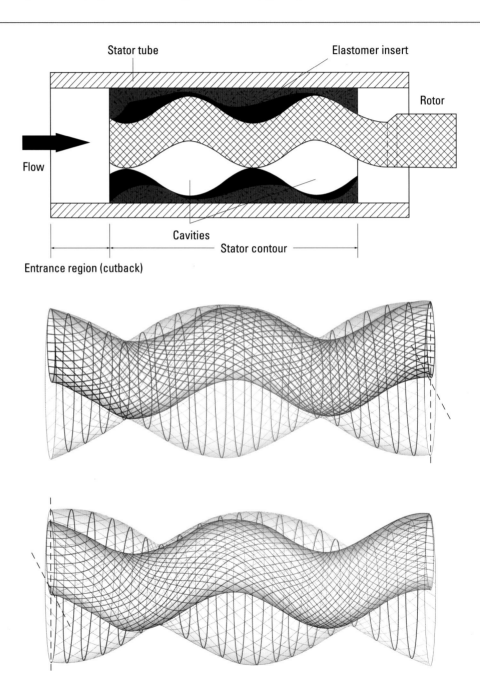

The stage length is defined as the axial length required for one lobe to rotate 360° along its helical path (Fig. 3-21). Stage length is also referred to as the "pitch."

The projected surface areas of the rotor in the radial and axial directions influence the amount of torque and axial thrust that a power section can create. The longer the stage length, the greater the projected surface area in the radial direction and the larger the force applied to turning the rotor. Generally, a longer stage length is more desirable because it affords more torque and less axial thrust in the power section. A longer stage length also results in lower rotational speed for a given lobe configuration. This characteristic is used in high-flow motors, which have long stage lengths assembled in long power sections with relatively few stages. This configuration allows higher flow rates, which generate lower rpm.

The hydraulic power extracted from the mud flow by the motor is

$$\textit{hydraulic power}\,(\text{hp}) = \textit{flow rate}\,(\text{galUS/min}) \times \textit{pressure drop}\,(\text{psi})/1{,}714, \qquad (3\text{-}1)$$

and the mechanical power output by a motor to the bit for drilling is

$$\textit{mechanical power}\,(\text{hp}) = \textit{rpm} \times \textit{torque}\,(\text{ft.lbf})/5{,}252, \qquad (3\text{-}2)$$

for which

- Mud flow rate is controlled at surface by adjusting the rig mud pump output.

- Pressure drop is controlled by the number of stages.

- The rpm and torque are controlled by the configuration of the motor. The greater the number of lobes and cavities, the higher the torque and the lower the rpm.

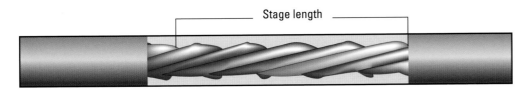

Stage length

Rotor

Figure 3-21. The stage length is the axial length required for a lobe to create one 360° spiral on the rotor.

The surface-adjustable bent housing allows the angle between the bit shaft and the axis of the motor to be changed, generally in the range from 0° to 3° (Fig. 3-22). Adjustment is performed only at the surface, where the stator housing can be unscrewed from the adjustment ring (1 and 2 in Fig. 3-22). The ring is then lifted to disengage the alignment teeth, and the ring can then be rotated to give the required bent housing angle (3). The adjustment ring is then slotted into position (4) and the stator adaptor screwed back into place to lock the ring (5).

The greater the bent housing angle, the greater the dogleg that can be achieved with the motor. However, high bent housing angles also cause significant stresses in the housing when rotary drilling, so the bend is kept as small as possible.

By orienting the surface-adjustable bent housing in the desired drilling direction and rotating the bit by pumping mud through the positive displacement motor, it is possible to drill in a desired direction. The process of keeping the drillstring oriented in a desired direction while drilling is called sliding. The orientation, or toolface, indicates the direction in which the well is to be deviated. In deviated and horizontal wells the toolface is referenced to the high side of the hole and is called the gravity toolface (GTF):

- 0° GTF indicates building angle.

- 90° GTF indicates turning right.

- 180° GTF indicates dropping angle.

- 270° GTF indicates turning left.

In nearly vertical holes the toolface is oriented relative to north. In this case it is called a magnetic toolface (MTF):

- 0° MTF indicates deviating north.

- 90° MTF indicates deviating east.

- 180° MTF indicates deviating south.

- 270° MTF indicates deviating west.

Because the angle on the bent housing is small (typically less than 3°), the steerable motor can also be rotated. This negates the effect of the bend and gives a relatively straight borehole, which is slightly over gauge (slightly larger than bit size). By alternating sliding and rotating intervals, the directional driller can control the rate at which the borehole angle is changed. The rate of change in borehole angle (or dogleg severity) is normally given in degrees per 100 ft [30 m].

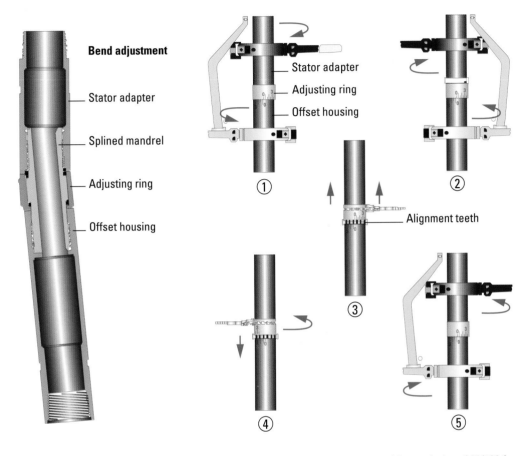

Bend adjustment

Stator adapter

Splined mandrel

Adjusting ring

Offset housing

Stator adapter

Adjusting ring

Offset housing

① ②

Alignment teeth

③

④ ⑤

Figure 3-22. The surface-adjustable bent housing allows changing the angle between the bit shaft and the axis of the motor.

For example, if a motor slides at 5°/100 ft and holds angle while rotating, then to achieve a dogleg of 2°/100 ft, the driller needs to slide for 2°/5° = 40% of the interval and rotate for the remaining 60%. If the interval to be drilled with a 2°/100-ft dogleg is 60 ft [18 m], then the driller slides for 40% × 60 ft = 24 ft [7 m] and rotates for the remaining 36 ft [11 m]. Depending on the intervals involved, the driller may chose to break the slide-rotate sequence into shorter intervals of sliding alternating with shorter intervals of rotating so that the overall change in well angle is the same but distributed more evenly along the 60 ft than if all the sliding is performed at the beginning.

Figure 3-23 shows a directional performance plot for a directional motor over numerous slide-rotate sequences. The darker background shading indicates slide sequences. The GTF (purple squares) is visible during the slide sections but not during the rotate sections because toolface is meaningless when the bent housing is rotating. The continuous azimuth (yellow squares connected by green line) overlies the static survey azimuth (purple line). The static survey inclination (yellow squares connected by blue lines) was taken every 90 ft [27 m] and indicates a relatively smooth increase in inclination up to about 59°, after which the inclination is held stable. The continuous inclination (red line) shows the well to be far more tortuous. During the slide sections the driller is building inclination, but the BHA drops inclination when rotating, which causes the slide-rotate sequence to result in an undulating borehole. The difference between the static surveys and the continuous inclination and azimuth data is not unusual. The averaging caused by the long distance between static survey stations results in a calculated trajectory that seems smoother than is really the case.

Figure 3-23. In this directional performance plot for a directional motor over numerous slide-rotate sequences, the oscillations in inclination (red line) caused by the slide-rotate sequence result in an undulating well trajectory.

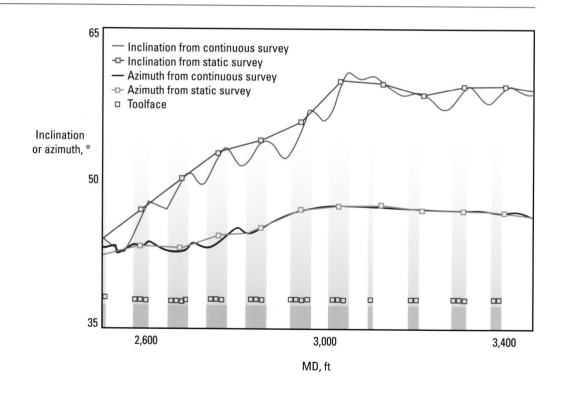

This effect is particularly significant for well placement using images. It is possible that the well trajectory calculated from the static surveys does not capture the true wellbore undulations to which the LWD images respond. For example, in a slide-rotate sequence the trajectory may start with an inclination of 90°, drop to 89°, intersect an underlying layer, and then build inclination back to 90° while drilling up out of the layer.

In this case the static surveys before and after the interval both show 90° so the trajectory is assumed to be horizontal. Because the LWD images show a layer being touched by the bottom of the well, the layer is assumed to dip up toward the trajectory, yielding an inaccurate dip estimation. This undulating trajectory caused by slide-rotate sequences increases the uncertainty on dips picked from LWD images in boreholes drilled with directional motors unless the continuous survey information is used to capture the true undulations of the borehole.

It should also be noted that over sliding intervals the LWD image data is unavailable not where the motor is sliding but over the interval where the LWD imaging sensors are sliding. This is due to the distance between the bit and the LWD imaging sensors.

Another difficulty with slide-rotate sequences is that orienting prior to each slide section is time consuming and reduces the overall rate of penetration achieved for the well. In addition, because the mud is not agitated during sliding, hole cleaning efficiency is reduced, and the probability of becoming differentially stuck as a result of the pressure imbalance between the wellbore and formation increases. Finally, the overall length of the well may be limited because static friction during sliding, which is greater than dynamic friction while rotating, may prevent effective weight transfer to the bit. To overcome these limitations, RSSs were developed.

3.3.5 Rotary steerable systems

Rotary steerable systems deliver continuous steering while rotating. Compared with drilling with a motor, continuous steering while rotating provides the following benefits:

- steadier deviation control—continuous rather than slide-rotate steering

- smoother in-gauge hole—no rotation of a bent housing causing hole enlargement

- better hole cleaning—cuttings are agitated into the mud flow by the rotation

- extended hole reach—reduced drag because rotating friction is less than static friction

- overall improvement in the rate of penetration—elimination of time spent orienting

- ensured acquisition of azimuthal formation data over the entire length of the well, which is particularly advantageous for well placement—rotation of the BHA is continuous.

The two main types of RSS are

- push-the-bit—applies side force to increase the side-cutting action of the bit

- point-the-bit—introduces an offset to the drilling trajectory similar to a bent housing but allowing continuous rotation.

A push-the-bit RSS uses pads on a bias unit to push against the borehole wall, which pushes the bit in the opposite direction (Fig. 3-24).

Figure 3-24. The three pads of a push-the-bit RSS push against the borehole wall to deflect drilling of the well in the opposite direction.

As the system rotates, the pads must be activated in sequence to ensure consistent steering in the desired direction. The control unit contains the electronics for control of the toolface and the percentage of time spent steering. The system operates by diverting a small percentage of the mud flow to activate the pads. By sensing the rotation of the BHA relative to the Earth's magnetic field and controlling an electric motor to rotate in the opposite direction, the control unit holds a control valve (blue element in Fig. 3-25) geostationary. The port in the control valve is oriented opposite the desired steering direction (toolface). Each of the three holes in the orange element in Fig. 3-25 guides mud behind one of the three pads. A small proportion of the mud flow is thus diverted behind each of the pads in sequence as the entry port to the piston behind each pad rotates in front of the geostationary port in the control valve. This causes the pads to open in sequence as they rotate into the "pushing" sector and apply force to the borehole wall opposite the desired steering direction. The pad opens a fraction of an inch because the diameter of the bias unit is only slightly smaller that the borehole diameter (Fig. 3-26). The drill bit, which must have cutting elements on the side (called side-cutting action) as well at the front, then preferentially cuts the rock on the side opposite the activated pad, resulting in a change in the well trajectory.

Because of the high power requirements of the control motor, the system has its own power-generation capabilities through a mud turbine and alternator assembly. In addition to supplying power to the motor to counter rotate the control valve against the rotation of the collar to keep it geostationary, the turbo-alternator system also supplies power to the electronics.

Steering is controlled by commanding the system to spend a certain percentage of the time steering (duty cycle) in a specific direction (toolface) and the remaining time in neutral. While in neutral the BHA could drill ahead straight if it is a hold assembly, but it could have build or drop tendencies depending on the configuration of the three touch points. The driller needs to take the BHA tendency into account when downlinking commands to the system.

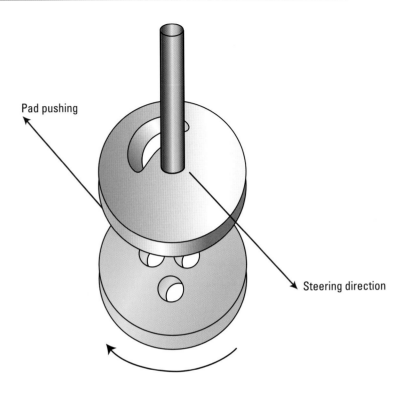

Pad pushing

Steering direction

Figure 3-25. The control valve (gray) is held geostationary by the control unit, which diverts a small proportion of the mud flow behind each of the pads in sequence as each pad's entry port rotates in front of the port in the control valve.

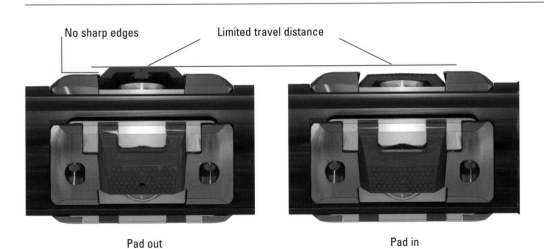

No sharp edges Limited travel distance

Pad out

Pad in

Figure 3-26. The diameter of the bias unit at the pads is only slightly smaller than bit size so the pads do not have to travel far before contacting and applying force to the borehole wall. Pad travel distance is limited to approximately ¾ in [1.9 cm].

Figure 3-27. Toolface and duty cycle options are programmed into the system before it is run in the hole. By downlinking to the system the driller can change to any other point of the steering command diagram for the RSS. The red point is an example discussed in the text.

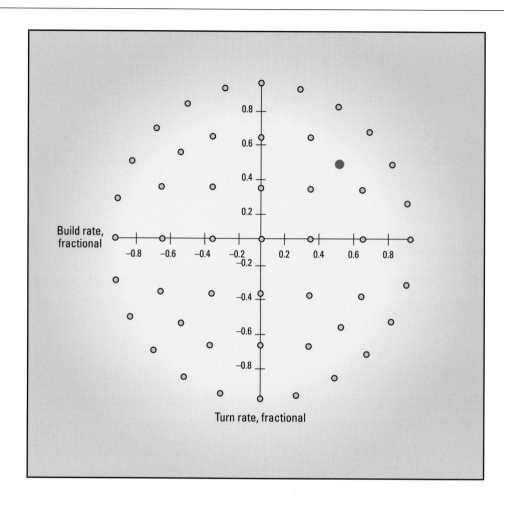

Downlinking is performed by adjusting the mud pumps at the surface to deliver a sequence of mud flow rate changes. The RSS detects these as changes in the rpm of the downhole turbine. Each point on the steering command diagram, an example of which is shown in Fig. 3-27, can be selected with a specific sequence of flow rate changes. Angles clockwise from the positive y-axis correspond to the GTF. The distance from the center corresponds to the proportion of time spent steering. For example, if the driller wants to build inclination as quickly as possible, the downlink sequence commanding the RSS to operate at the point at the top of the diagram would be sent. This corresponds to 100% duty cycle at a GTF of zero.

Downlinking to the RSS to operate at the red point in the top right quadrant would result in the RSS spending 67% of drilling time steering up and to the right. This should result in the trajectory building and turning to the right; however, depending on the interaction of the RSS with the formation, the turn rates vertically and horizontally may be different. If the BHA has a strong drop tendency, this amount of build could be required just to maintain the well at horizontal.

The downlink command to the RSS is the means by which the system performance is adjusted. The interaction of the entire BHA with the formation then determines the trajectory. Many systems now incorporate an inclination hold feature. In this mode of operation the driller downlinks the desired borehole inclination, and the RSS automatically adjusts the toolface and duty cycle settings to maintain the requested inclination. This mode of operation can deliver very straight trajectories and is particularly useful for well placement with deep directional measurements; for example, the well can be drilled parallel to a remotely detected boundary.

Because of the requirement to push off the opposite side of the borehole to cause a change in the trajectory, push-the-bit systems are sensitive to the mechanical properties of the formation. Kicking off from vertical in an existing well can also be problematic because the pads become unable to contact the borehole wall at the hole enlargement that occurs at the kickoff point. Point-the-bit RSSs overcome these limitations.

A point-the-bit RSS delivers all the benefits of a push-the-bit system with reduced sensitivity to the formation, resulting in more consistent steering, and generally higher dogleg capability. The point-the-bit system is centered on a universal joint that transmits torque and weight on bit but allows the axis of the bit to be offset with the axis of the system. The axis of the bit is kept offset by a mandrel maintained in a geostationary orientation through the use of a counter-rotating electrical motor (Fig. 3-28). Whereas the push-the-bit system keeps a control valve geostationary to divert mud behind the pads, the point-the-bit system keeps a mandrel geostationary.

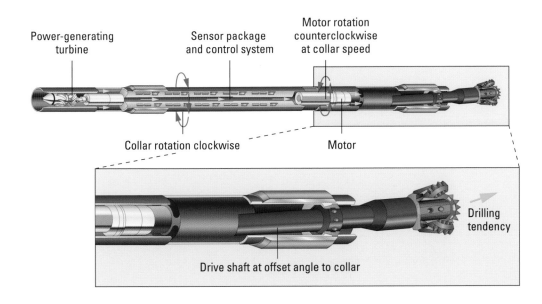

Power-generating turbine

Sensor package and control system

Motor rotation counterclockwise at collar speed

Collar rotation clockwise

Motor

Drive shaft at offset angle to collar

Drilling tendency

Figure 3-28. An electric motor is used by the PowerDrive Xceed point-the-bit system to counter rotate a mandrel against the rotation of the collar. This keeps the mandrel, and thus the bit, oriented in the same direction while still rotating with the collar

Figure 3-29. A geostationary offset angle between the bit and collar creates the steering tendency for the PowerDrive Xceed point-the-bit system.

Similar to a push-the-bit RSS, high power requirements require that the system have its own power-generation capabilities through a high-power turbine and alternator assembly. The system also contains power electronics to control the motor and sensors that monitor the rotation of the collar and motor (Fig. 3-29). The sensors provide input and feedback for the control of the system.

As with the push-the-bit system, the directional driller controls the dogleg by downlinking to the system to change the proportion of steering versus neutral time.

Because there is no rotation provided downhole by either type of RSS, the entire drillstring must be continuously rotated from surface. If additional downhole rpm is desired or surface rotation must be kept to a minimum (such as when casing wear is a concern), a mud motor (without the bent housing) can be used above the RSS to provide downhole rotation of the RSS assembly.

3.4 Landing calculations

Although the majority of trajectory calculations are performed by the drilling engineer and directional driller, understanding how to make a quick calculation of well location can be useful (Fig. 3-30). The following calculations are performed in feet, but the same logic can be applied to derive the results in meters. Note that this calculation considers change in inclination only. There is no change in azimuth.

Consider drilling a circle of radius, R, and circumference, C, at a continuous build rate, $B = 1°/100$ ft (Fig. 3-30). To cover the $360°$ encompassed by the circle, the circumference is

$$C = 360° \times B \qquad \qquad (3\text{-}3)$$
$$= 360° \times 100 \text{ ft}/1° = 36,000 \text{ ft.}$$

This value is substituted for C in the following equation to calculate the radius of a vertical circle drilled at the specified continuous build rate:

$$C = 2\pi R, \qquad \qquad (3\text{-}4)$$

which can be restated as

$$\Longrightarrow \quad R = C/2\pi$$
$$= 36,000 \text{ ft}/2\pi$$
$$= 5,730 \text{ ft for a buildup rate of } 1°/100 \text{ ft}$$

or

$$\Longrightarrow \quad R = 5,730 \text{ ft}/B. \qquad \qquad (3\text{-}5)$$

This simple relationship allows calculation of the well position.

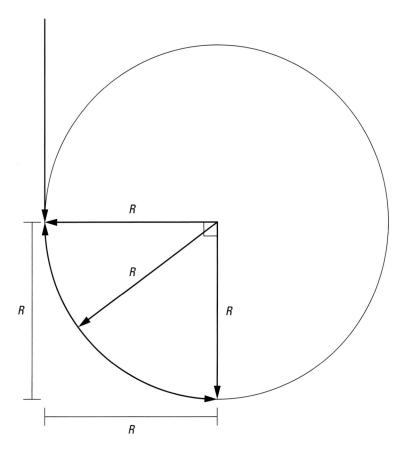

Consider the problem of a well being landed at horizontal (90° inclination) just below the top of the reservoir (Fig. 3-31).

The well is at 87° inclination when 2 ft above the top of the reservoir. The plan is to land the well at 90°, 4 ft below the top of the reservoir. If the remaining section of well is drilled at 3°/100 ft, will the well be too high or too low?

From the relationship between the build rate and radius of curvature, the radius can be calculated:

$R = 5{,}730 \text{ ft}/B$
$\quad = 5{,}730 \text{ ft}/3$
$\quad = 1{,}910 \text{ ft}.$

The change in inclination from 87° to 90° is 3°.

Figure 3-31. In this example well landing problem, the well has an inclination of 87° at 2 ft above the top of the reservoir and the plan is to land the well at 90° inclination and 4 ft below the top of the reservoir.

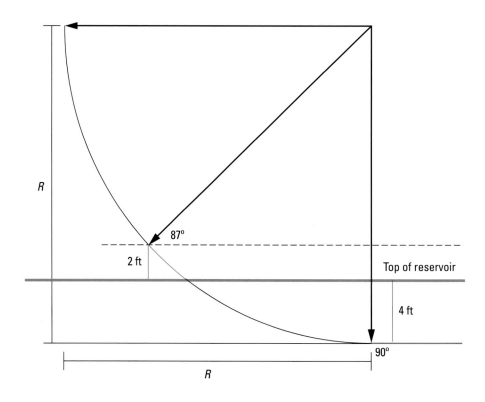

As shown in Fig. 3-32 the distance between the center of curvature and the current TVD of the well (dashed horizontal line in Fig. 3-32) can be calculated from the right-angle triangle it forms with the radius to the current well location and the current well TVD.

The vertical distance between the center of curvature and the current TVD of the well is

$R \times \cos 3° = 1{,}910 \text{ ft} \times 0.9986$
$\qquad = 1{,}907.4 \text{ ft}.$

The TVD change from the current well location to where it will land at a continuous build rate of 3° is R minus this distance:

$TVD\ change = 1{,}910 \text{ ft} - 1{,}907.4 \text{ ft}$
$\qquad = 2.6 \text{ ft}.$

In this example the TVD change from the current well location is 2.6 ft. Given that the well is currently 2 ft above the top of the reservoir, this means that the well will land only 0.6 ft below the top if the 3°/100 ft build rate is maintained. The well will land too high.

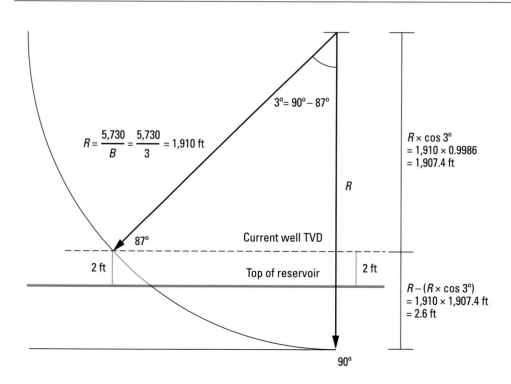

$3° = 90° - 87°$

$R = \dfrac{5{,}730}{B} = \dfrac{5{,}730}{3} = 1{,}910 \text{ ft}$

87°

Current well TVD

2 ft

Top of reservoir

R

90°

$R \times \cos 3°$
$= 1{,}910 \times 0.9986$
$= 1{,}907.4 \text{ ft}$

2 ft

$R - (R \times \cos 3°)$
$= 1{,}910 \times 1{,}907.4 \text{ ft}$
$= 2.6 \text{ ft}$

Figure 3-32. Will the well land too high or too low? Simple trigonometric calculations show that it will land too high, just 0.6 ft below the top of the reservoir.

The plan is to land 4 ft below the top of the reservoir. The build rate required to achieve the landing can be calculated by first solving for the radius, as shown in Fig. 3-33.

The vertical radius can be expressed as two sections:

- $R \times \cos 3°$ above the current TVD of the well

- $R - (R \times \cos 3°)$ below the current TVD of the well.

The plan calls for the well to land 6 ft below the current TVD of the well, so the lower section of the radius must equal 6 ft:

$$R - (R \times \cos 3°) = 6 \text{ ft}$$
$$R(1 - \cos 3°) = 6 \text{ ft}$$
$$R = 6 \text{ ft}/0.00137$$
$$= 4{,}378 \text{ ft}$$

Now the value for R can be substituted into $R = 5{,}730 \text{ ft}/B$:

$$B = 5{,}730 \text{ ft}/R$$
$$= 5{,}730 \text{ ft}/4{,}378 \text{ ft}$$
$$= 1.23°/100 \text{ ft build rate required to land the well on target.}$$

Figure 3-33. The vertical radius is expressed as two sections to solve for the dogleg required to land the well in the correct position.

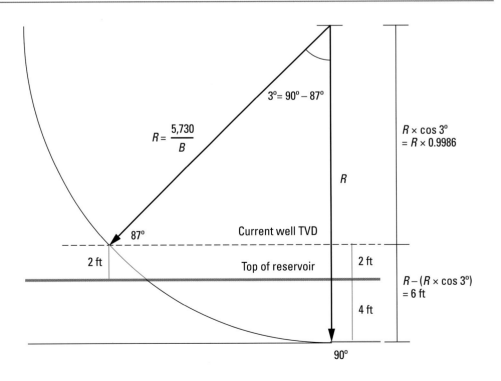

3.5 Communication with the directional driller

Clear communication of target changes to wellsite personnel is one of the most important aspects of well placement. Communication of trajectory changes should have the following qualities.

- Possible—Each BHA has capabilities and limitations that should be taken into account. There is no point asking for a dogleg severity that the BHA is unable to deliver.

- Simple—The simpler the description of what is required, the more likely it will be achieved. Communication should be kept simple and concise.

- Quantitative—Requested changes should be clearly stated using numbers to define the required TVD and well inclination at that TVD.

  - Where a target point within the reservoir is specified it is best to define the TVD and well inclination required at the target point. It is not sufficient to request a change in TVD without specifying the inclination required at that TVD.

  - Where the well is steered to follow a boundary or stay within a formation of known dip, the required well inclination should be communicated to the driller.

- Based upon clearly understood LWD tool responses—The only reason to change the target is because of changes in the geology. There should be a clear understanding (preferably a consensus) of what the log responses mean and the appropriate course of action. Too often rapid changes in well placement instructions are given because of initial misinterpretation of a tool response, resulting in an undulating borehole and a frustrated driller.

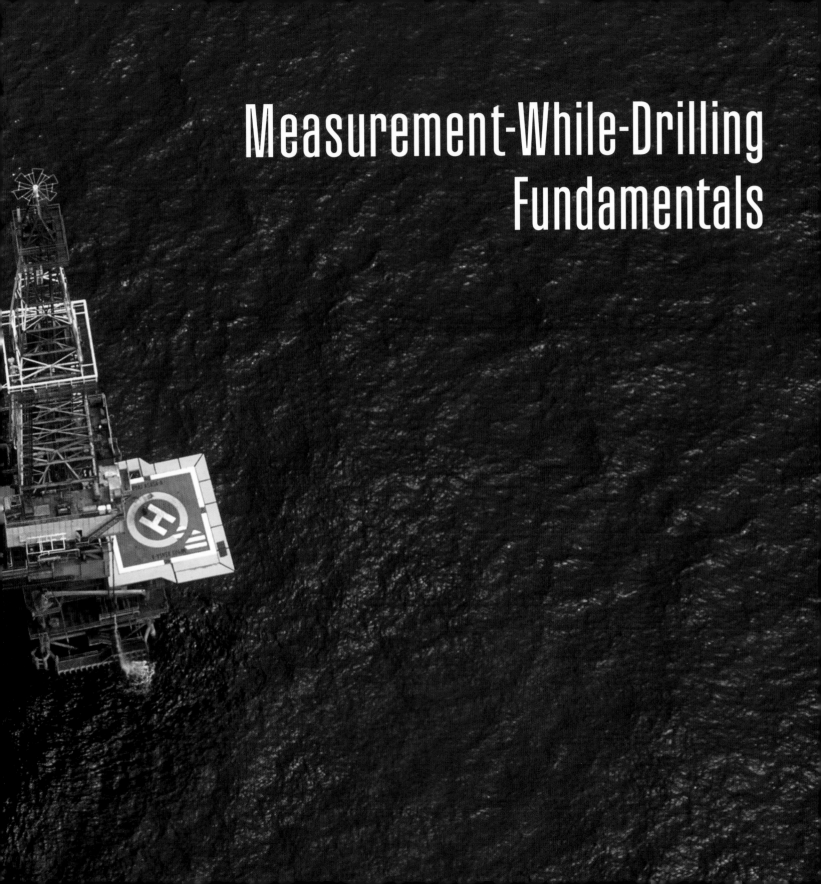

Measurement-While-Drilling
Fundamentals

Measurement while drilling (MWD) is the evaluation of physical properties, usually including pressure, temperature, and wellbore trajectory in 3D space, while extending a wellbore.

4.1 Definition

MWD is now standard practice in most directional wells, where the tool cost is offset by rig time savings and wellbore stability improvements. The measurements are made downhole, transmitted to the surface, and also stored in solid-state memory for retrieval once the tool has returned to the surface. Real-time data transmission methods vary from company to company but usually involve digitally encoding data and transmitting it to the surface as pressure pulses in the mud system. The pressure pulses can be positive, negative, or continuous sine waves. Some MWD tools have the ability to store measurements for later retrieval with wireline or when the tool is tripped out of the hole if the data transmission link fails. MWD tools that measure formation parameters (resistivity, porosity, sonic velocity, and GR) are referred to as logging-while-drilling (LWD) tools. LWD tools use similar data storage and transmission systems, and some have more solid-state memory to provide higher resolution logs for retrieval after the tool is tripped out than is possible with the relatively low-bandwidth, mud-pulse data transmission system.

The following sections in this chapter review the four major capabilities of typical MWD tools:

- real-time surveys for directional control—inclination, azimuth, and toolface

- real-time power generation

- real-time mud-pulse data transmission telemetry system

- real-time drilling-related measurements—weight on bit, torque at bit, and mud pressure.

4.2 BHA orientation: Toolface

When drilling directionally, the orientation of the drilling system defines the direction in which the well deviates. Toolface is the angle between a reference, either gravity in a deviated well or the north reference (true, magnetic, or grid north) in a vertical well, and the direction in which the BHA tends to deviate the hole. In the case of drilling a deviated well with a surface-adjustable bent housing on a positive displacement motor, toolface is the angle between the bend in the housing and the high side of the hole. The toolface measurement is transmitted to the surface in real time so the directional driller can ensure that the drilling assembly is oriented to give the desired direction to the well.

Toolface is used during the drilling of a well. Once the well has been drilled, the well inclination and azimuth define the location of the borehole in 3D space.

4.2.1 Magnetic toolface

Magnetic toolface is used in vertical and nearly vertical wells up to about 3° inclination to define the azimuth relative to north in which the well deviates as drilling proceeds.

The magnetometers in the survey tool are used in conjunction with the accelerometers to determine magnetic north and subsequently true north. The BHA is then oriented relative to true north and the MTF transmitted in real time so the directional driller can make adjustments if required prior to drilling.

The MTF is stated clockwise from north, so an MTF of zero indicates that the well will be drilled due north, whereas an MTF of 90° indicates that the well will be drilled east. In the example shown in Fig. 4-1, the MTF indicates that the well will be drilled to the southwest.

4.2.2 Gravity toolface

Once a well has been kicked off in a required azimuth by using the MTF measurement, the toolface reference changes to the high side or top of the hole. The reference is changed because in a deviated well it is possible to increase and decrease inclination and turn left and right, which cannot be easily referenced to north.

The gravity toolface measurement is used in deviated and horizontal wells to define the direction in which the borehole deviates relative to the top of the hole. The accelerometers in the survey tool are used to define the top of the hole and the GTF is measured clockwise from the top of the hole, looking down the hole in the direction of drilling. A GTF of zero indicates that the well will increase inclination, whereas a GTF of 90° indicates that the well will drill to the right. In the example shown in Fig. 4-2, the GTF indicates that the well will both build angle and turn to the right.

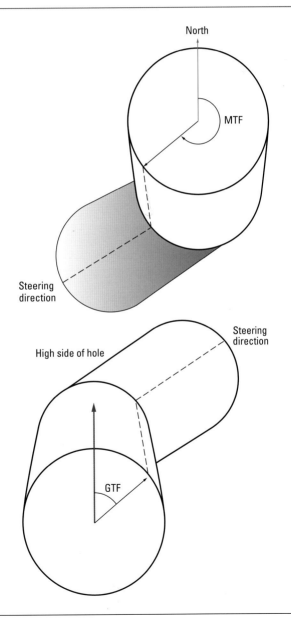

Figure 4-1. Magnetic toolface is used in vertical or nearly vertical wells to define the azimuth relative to north of the steering direction.

Figure 4-2. Gravity toolface is used in deviated and horizontal wells to define the direction in which the borehole deviates relative to the top of the hole.

4.3 Wellbore surveying

Surveying is used to define the position of a well in 3D space. Survey measurements taken at discrete points along the wellbore length are used to calculate the position of the wellbore in 3D space (Fig. 4-3).

Three fundamental measurements define a survey point:

- along-hole or measured depth at which a survey is taken
- wellbore inclination relative to the Earth's gravitational field
- wellbore azimuth relative to north.

4.3.1 Measured depth

The measured depth of a point in a well as defined by the driller is the cumulative length of drillpipe that has been run in the hole to position the bit at that point. To make this measurement, each piece or joint of drillpipe is manually measured at surface with a tape measure and the specific length of each piece of pipe added to

Figure 4-3. The trajectory of the well (green line) is calculated from the survey measurements taken at discrete points (red dots) to define the position of a well in 3D space.

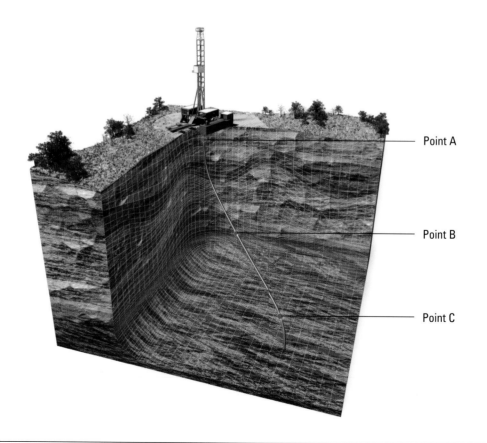

Point A

Point B

Point C

the tally as the pipe is run in the hole. No corrections for temperature, pressure, stretch, or compression of the drillpipe are applied.

MWD depth is measured continuously by using a device that determines how far the traveling block holding the drillpipe moves during drilling. At the end of drilling each joint, or several joints (known as a stand) of drillpipe, the MWD depth measurement is adjusted to agree with the driller's depth. This generally requires applying a very small change, which is linearly distributed back along the length of the joint or stand.

4.3.2 Inclination

Inclination is the angle between a vertical line and the path of the wellbore at that point (Fig. 4-4). The vertical reference is the Earth's gravitational field because it always points to the center of the Earth. An inclination of zero indicates a vertical well. An inclination of 90° indicates a horizontal well. This is opposite to the convention for formation dip, for which a dip of zero indicates a horizontal layer. Caution is necessary because drillers reference their angles from vertical whereas geologists reference their angles from horizontal.

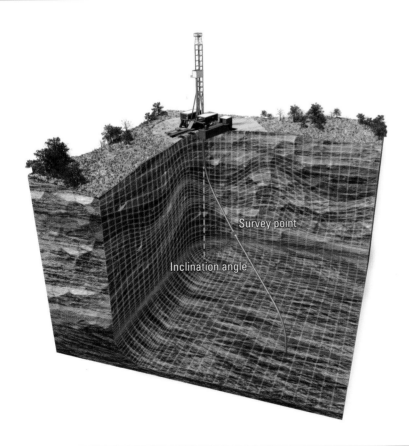

Figure 4-4. Inclination is the angle between a vertical line and the path of the wellbore at that point.

Inclination is measured using triaxial accelerometers in the MWD tool to determine the three components of the Earth's gravity vector relative to the position of the tool (Fig. 4-5).

The vector sum of the three measurements equals the total gravity vector, $\mathbf{G}$:

$$\mathbf{G} = \sqrt{\mathbf{G}_x^2 + \mathbf{G}_y^2 + \mathbf{G}_z^2},$$

(4-1)

where

$\mathbf{G}_x$ = component of the gravity vector measured along the axis of the tool
$\mathbf{G}_y$ = component of the gravity vector measured perpendicular to the axis of the tool
$\mathbf{G}_z$ = component of the gravity vector measured perpendicular to both the axis of the tool and the y axis of the tool.

The vector sum is compared with the expected gravitational field value at the specified location on the face of the Earth to ensure that the three accelerometers are functioning correctly. If one of the accelerometer readings has drifted, the vector sum does not equal the local gravity value.

Figure 4-5. Inclination is measured using three orthogonal accelerometers in the MWD tool to measure the Earth's gravitational field.

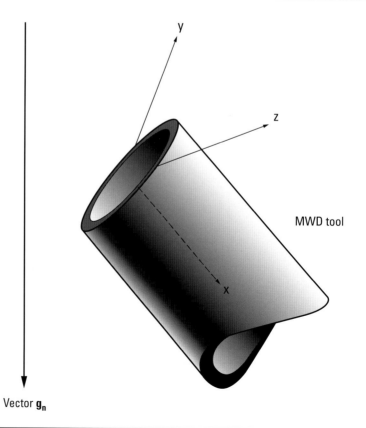

Vector $\mathbf{g}_n$

To calculate the inclination, only the component along the axis of the tool (the x axis) is required. $\mathbf{T_g}$ is the vector sum of the $\mathbf{G_y}$ and $\mathbf{G_z}$ components (as shown in Fig. 4-6, the components perpendicular to the axis of the tool):

$$inclination = \cos^{-1}\left(\frac{\mathbf{G_x}}{\mathbf{G}}\right).$$

(4-2)

When the tool is vertical the x-axis accelerometer measures the entire gravity vector, so the ratio $\mathbf{G_x}/\mathbf{G} = 1$, which gives an inclination value of zero. If the tool is horizontal then the x-axis is perpendicular to the gravity vector and the x-accelerometer reads zero, for which the ratio $\mathbf{G_x}/\mathbf{G} = 0$ gives an inclination of 90°.

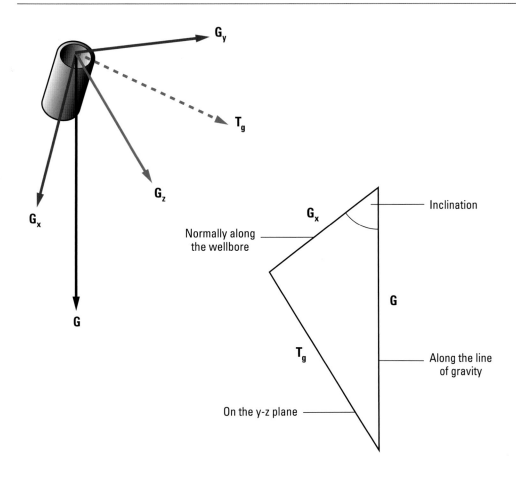

Figure 4-6. The inclination is determined from the component of the gravitational vector along the centerline axis of the tool.

4.3.3 Azimuth

Azimuth is the angle between the north reference and a horizontal projection of the current survey position (Fig. 4-7). To determine the azimuth of a well at a survey point, the MWD tool must measure the Earth's magnetic field, which then allows determining the direction of the north reference. An azimuth of zero indicates that the well is drilling in the direction of the north reference. An azimuth of 90° indicates that the well is drilling east relative to the north reference.

Figure 4-7. Azimuth is the angle between the north reference and a horizontal projection of the current survey position.

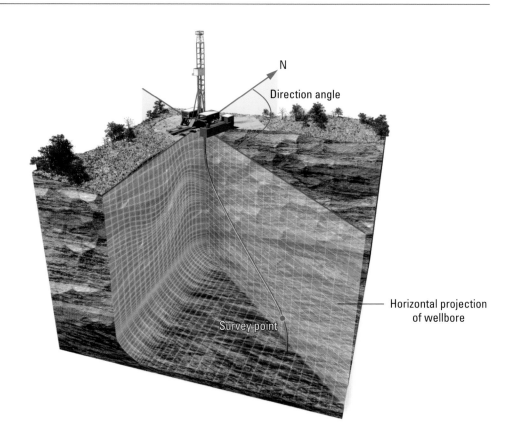

N

Direction angle

Survey point

Horizontal projection
of wellbore

The calculation of azimuth is more complicated because the Earth's magnetic field changes direction across the face of the Earth (Fig. 4-8). In addition, the magnetic field direction also changes with time as the axis of the core of the Earth (magnetic north-south axis), the spinning of which creates the magnetic field, drifts relative to the axis around which the remainder of the planet rotates (true north-south axis).

Magnetic declination is the angle between magnetic and true north at a given point on the Earth.

Magnetic dip is the angle between the magnetic flux lines and a line tangent to the Earth's surface. The magnetic dip is +90° at the north magnetic pole, −90° at the south magnetic pole, and zero close to the equator.

Triaxial magnetometers are used to measure the components of the Earth's magnetic field along and perpendicular to the MWD tool. The vector sum of the three measurements equals the total magnetic vector, **H**:

$$\mathbf{H} = \sqrt{\mathbf{H}_x^2 + \mathbf{H}_y^2 + \mathbf{H}_z^2}.$$

(4-3)

This value is compared with the expected magnetic field strength at that time and place on the Earth to confirm the correct functioning of the three orthogonal magnetometers.

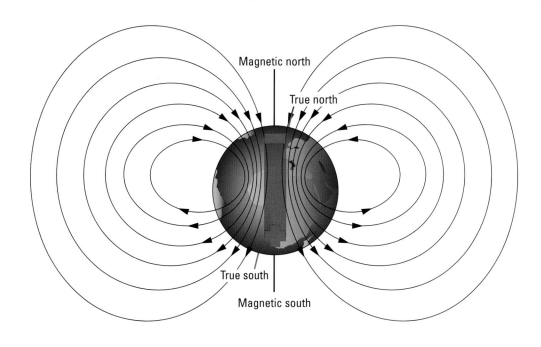

Figure 4-8. The Earth's magnetic field changes in strength and direction depending where and when it is measured.

To calculate the well azimuth, the horizontal component of the magnetic flux line is used to define the direction of magnetic north (Fig. 4-9):

$$horizontal\ component = magnetic\ field\ strength \times cos(magnetic\ dip). \qquad (4\text{-}4)$$

Remember that the MWD tool is measuring the three magnetic components from its position in the deviated borehole. To be able to define the horizontal plane, the inclination of the MWD tool must be known. The inclination data derived from the accelerometers is used to define the inclination rotation back to horizontal. Hence, for calculations of azimuth the accelerometers and magnetometers are used.

Figure 4-9. The horizontal component of the magnetic flux line is used to define the direction to magnetic north.

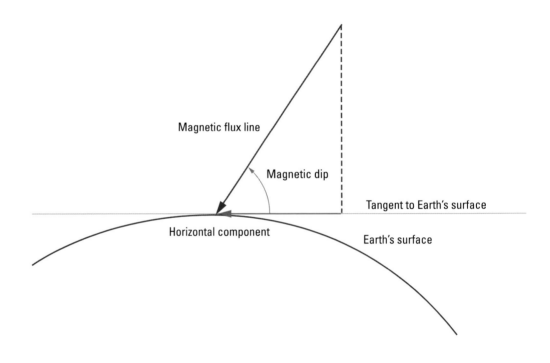

Although the tool measures the azimuth with respect to magnetic north, oil fields are typically mapped against a local grid that may be slightly rotated, even from true north. There are three possible north references (Fig. 4-10):

- true—azimuth measured with reference to true north

- magnetic—azimuth measured with reference to magnetic north

- grid—azimuth measured with reference to grid north (the top of the map, depending upon the cartographic projection).

To convert from one reference to the other, the angles between them must be taken into account:

- magnetic declination—angle D between magnetic and true north

- grid convergence—angle C between grid and true north

- total correction = $D - C$.

Surveys are typically referenced to grid north instead of true north, specifically so the directional driller can determine whether the well is on plan and within the lease boundaries.

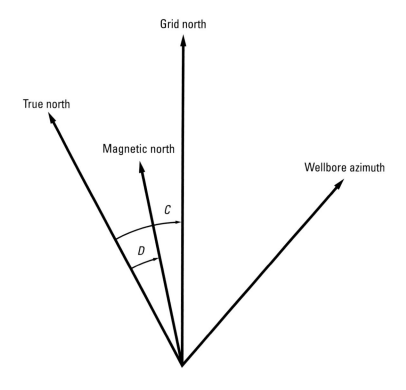

4.3.4 Well trajectory calculations

To describe the well trajectory, the coordinates x, y, and z must be determined from the MD, inclination, and azimuth at discrete survey points along the well path. The coordinates x and y represent the departures in the west-east and south-north directions, respectively, and z is the vertical departure. Of the several calculation methods available, the most accurate and common is the minimum curvature method.

The minimum curvature method uses the angles measured at two consecutive survey points to describe a smooth, circular curve that represents the wellbore path (Fig. 4-11).

The computed wellbore curvature is usually expressed as a rate of change in the well angle (both inclination and azimuth) per 100 ft [30 m]. Called the dogleg severity, this represents the tortuosity of the well path.

Between each survey pair we calculate Δx, Δy, and ΔTVD from the MD, inclination, and azimuth at the survey points. Summing Δx, Δy, and ΔTVD defines the position of the wellbore in 3D space.

Figure 4-11. The minimum curvature method solves for the straightest smooth arc that can connect two survey points while respecting the MD between them and the inclination (*I*) and azimuth (*A*) measured at each of the survey points.

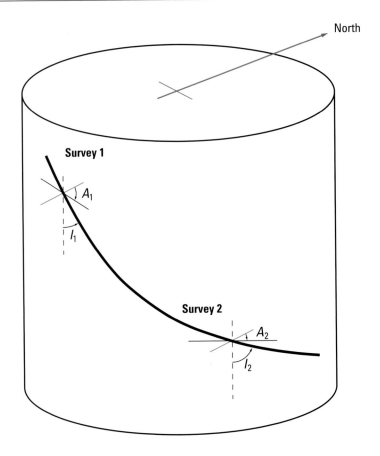

4.3.5 Survey report

A typical survey report is shown in Fig. 4-12. The first page includes all the data required to identify the well, the surveyed interval, trajectory calculation methods, and the various elevation, gravitational, and magnetic references used. Note that the gravitational and magnetic field strengths are in units of mGal (one thousandth of the nominal gravitational acceleration) and HCNT (H counts, one thousandth of Earth's nominal magnetic field strength), respectively. In the lower right panel of the first survey page the magnetic declination and grid convergence numbers are listed, along with the total azimuth correction applied to this survey, which is referenced to grid north because both corrections are applied (see Section 4.3.3, "Azimuth").

```
                          SCHLUMBERGER

                   Survey report      3-Jul-2009 22:32:13      Page   1 of 2

Client..................: Example
Field...................: Gusher

Well....................: Gusher #1           Spud date...............: 06-Jun-09
Max BHT.................: 243 deg F           Last survey date.........: 03-Jul-09
Engineer................: Tom                 Total accepted surveys...: 31
                                             MD of first survey.......: 13255.00 ft
Rig:....................: Rig #5              MD of last survey........: 16224.00 ft
State:..................: Big Rock

----- Survey calculation methods-------------   ----- Geomagnetic data ----------------
Method for positions.....: Minimum curvature    Magnetic model...........: BGGM version 2009
Method for DLS...........: Mason & Taylor       Magnetic date............: 25-Jun-2009
                                                Magnetic field strength..:   859.80 HCNT
----- Depth reference -----------------------   Magnetic dec (+E/W-).....:     1.47 degrees
Permanent datum..........: Mean Sea Level       Magnetic dip.............:    37.02 degrees
Depth reference..........: Driller's Depth
GL above permanent.......:     9.00 ft          ----- MWD survey Reference Criteria ---------
KB above permanent.......:    44.00 ft          Reference G..............:   998.88 mGal
DF above permanent.......:    44.00 ft          Reference H..............:   859.80 HCNT
                                                Reference Dip............:    37.02 degrees
----- Vertical section origin----------------   Tolerance of G...........: (+/-) 2.50 mGal
Latitude (+N/S-).........:     0.00 ft          Tolerance of H...........: (+/-) 6.00 HCNT
Departure (+E/W-)........:     0.00 ft          Tolerance of Dip.........: (+/-) 0.45 degrees

----- Platform reference point--------------    ----- Corrections ------------------------
Latitude (+N/S-).........:     n/a              Magnetic dec (+E/W-).....:     1.47 degrees
Departure (+E/W-)........:     n/a              Grid convergence (+E/W-).:    -1.25 degrees
                                                Total az corr (+E/W-)....:     2.72 degrees
Azimuth from Vsect Origin to target:  50.43 degrees   (Total az corr = magnetic dec - grid conv)
                                                Survey Correction Type ...:
                                                  I=Sag Corrected Inclination
                                                  M=Schlumberger Magnetic Correction
                                                  S=Shell Magnetic Correction
                                                  F=Failed Axis Correction
                                                  R=Magnetic Resonance Tool Correction
                                                  D=Dmag Magnetic Correction

SCHLUMBERGER Survey Report                3-Jul-2009 22:32:13        Page   2 of 2

===  ========  ======  =======  ======  ========  ========  ================  ================  =====  ===== ======
Seq  Measured  Incl    Azimuth  Course  TVD       Vertical  Displ    Displ    Total    At       DLS    Srvy  Tool
#    depth     angle   angle    length  depth     section   +N/S-    +E/W-    displ    Azim     (deg/  tool  Corr
-    (ft)      (deg)   (deg)    (ft)    (ft)      (ft)      (ft)     (ft)     (ft)     (deg)    100f)  type  (deg)
===  ========  ======  =======  ======  ========  ========  ================  ================  =====  ===== ======
```

Figure 4-12. The information on the first page of a survey report identifies the well, surveyed interval, trajectory calculation methods, and references. The subsequent pages report the measured and calculated well location.

The subsequent pages of the survey report list the measured and calculated well location information. The columns are as follows.

- Sequence number—Each row is numbered corresponding to the accepted survey number.

- Measured depth—MD at the D&I component in the MWD tool is specified.

- Inclination angle—This is the well inclination. If corrections have been applied, they are specified in the last column.

- Azimuth angle—This is the well azimuth. If corrections have been applied, they are specified in the last column. The azimuth reference is determined by which corrections have been applied. If only the declination correction has been applied, the azimuth is referenced to true north. If the grid convergence correction has also been applied, then the azimuth is referenced to grid north.

- Course length—The difference in MD is between the current survey and the previous survey. The closer the surveys are together, the more accurate the representation of the position of the borehole. Widely spaced surveys cannot capture the undulations in the wellbore and hence cannot deliver an accurate borehole location.

- TVD—The true vertical depth at this survey point is computed from the cumulative sum of all the ΔTVD increments computed between the surveys to this depth. The first point in this example identifies the survey tool type (second-to-last column) as TIP (for tie-in point), which means that all the data for this line, including the TVD, has been entered manually from another source.

- Vertical section—This is the projection of the straight-line horizontal distance between the origin (usually vertically below the wellhead) and the survey point onto a specified azimuth (Fig. 4-13). The azimuth from the vertical section origin to the survey point is specified at the bottom of the left column on the first page of the survey report.

- Displacement +N/S– —This component of the survey point horizontal displacement is from the origin along the north-south axis. Displacement to the north is positive.

- Displacement +E/W– —This component of the survey point horizontal displacement is from the origin along the east-west axis. Displacement to the east is positive.

- Total displacement—This is the minimum straight-line horizontal distance between the origin (usually vertically below the wellhead) and the survey point. The azimuth along which this total displacement is measured must also be specified because it changes as the well turns.

- At azimuth—The azimuth from the origin to the survey point is the azimuth along which the total displacement is measured.

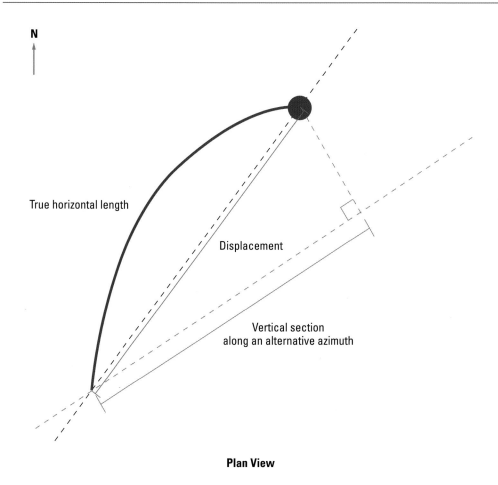

N

True horizontal length

Displacement

Vertical section
along an alternative azimuth

Plan View

Figure 4-13. Plan view shows displacement as a straight horizontal line between the origin and the survey point. A vertical section is the projection of the displacement onto a specified azimuth.

- DLS—The dogleg severity is calculated between the present and previous survey points. DLS indicates how sharply the well is changing direction (both inclination and azimuth).

- Survey tool type—The origin of the measured data is indicated. The first survey point in the example in Fig. 4-12 lists TIP for tie-in point, indicating that the driller has been given the data on the first line from which to start the trajectory calculations. This may be due to sidetracking out of an existing well.

- Tool corrections—Corrections that have been applied to the surveys are indicated in the last column. The possible types of correction are outlined in the lower part of the right column on page one.

4.3.6 The ellipsoid of uncertainty

Every measurement has an associated uncertainty. Conversion of the measured depth, inclination, and azimuth to 3D coordinates must take into account the uncertainty in each of the measurements and in the assumptions made in calculations (for example, that the well follows the path of minimum curvature between the survey points). Because the x, y, and TVD positions of a survey are computed from the sum of all the previous Δx, Δy, and ΔTVD increments computed between survey points, the uncertainty in the position of the deepest survey includes the cumulative uncertainty of all the surveys to that point.

There are numerous sources of uncertainty in depth, such as drillpipe length measurement uncertainty, pipe stretch, thermal expansion, and pressure effects. Inclination and azimuth uncertainty may accumulate from inherent instrument accuracy limitations, instrument alignment errors in the tool, as well as tool misalignment and sag in the borehole. Magnetic interference from the drillstring and external sources adds further uncertainty to the measured azimuth.

A number of mathematical models can be used to estimate the uncertainty in the wellbore position. The most recent, from the Industry Steering Committee on Wellbore Survey Accuracy (ISCWSA),[†] relies on a mathematical description of all error sources for all well types, locations, and tool performances. The output from this mathematical model is a 3D description of the probability distribution of the well location. Because the position of the wellbore is defined in 3D space, the uncertainty of the wellbore position is a 3D problem (Fig. 4-14). This 3D volume of uncertainty is called the ellipsoid of uncertainty.

The size of the ellipsoid of uncertainty is specified along the TVD and the semimajor and semiminor axes (Fig. 4-15). The ellipse of uncertainty is the calculated positional uncertainty based on the cumulative survey uncertainties at each survey station.

Factors that affect the size of the ellipsoid of uncertainty include

- survey tool types used—single-shot, MWD, gyroscope

- survey frequency—how often the well is surveyed

- BHA configuration—stabilizer placement, BHA sag

- latitude and longitude—location at the Earth's surface.

[†]Williamson, H.: "Accuracy Prediction for Directional MWD," paper SPE 56702 presented at the SPE Annual Technical Conference and Exhibition, Houston, Texas, USA (October 3–6, 1999).

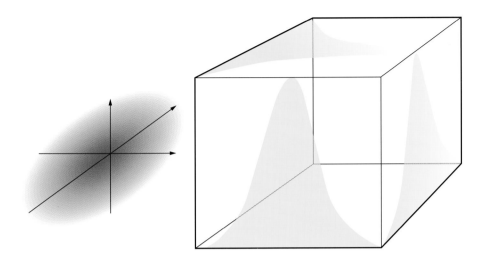

Figure 4-14. Three-dimensional uncertainty (in this case, normal distributions in each dimension) creates an ellipsoid of uncertainty around the computed well location.

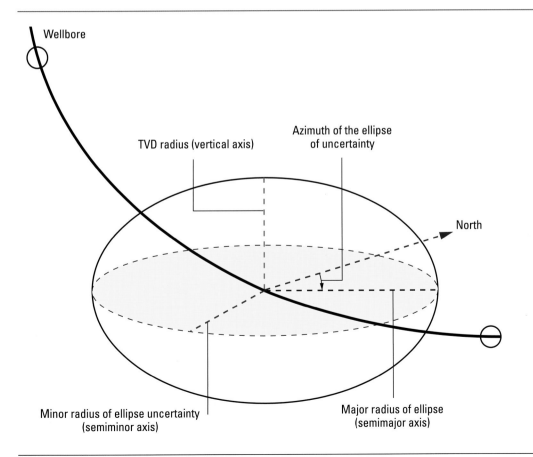

Figure 4-15. The ellipsoid of uncertainty defines the 3D volume with a specific probability of containing the wellbore trajectory.

Because each survey point has its own ellipsoid of uncertainty that takes into account the cumulative uncertainties from the previous survey points, the ellipsoids grow progressively larger as the well depth increases (Fig. 4-16).

Figure 4-16. The ellipsoid of uncertainty at each depth includes the cumulative uncertainty of all the previous surveys.

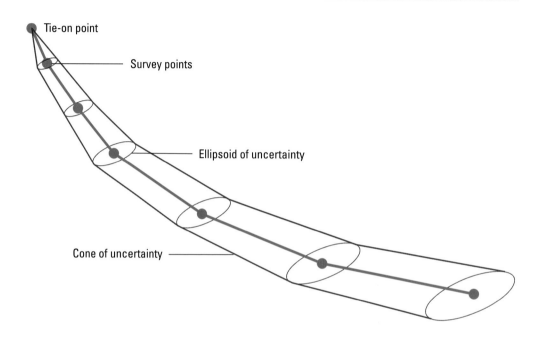

Tie-on point

Survey points

Ellipsoid of uncertainty

Cone of uncertainty

An alternative representation of the uncertainty is in the form of a cone of uncertainty around the wellbore (Fig. 4-17).

While the nominal well location (bright pink line inside the cone) may be in the zone of interest, if the uncertainty on the well position is larger than the thickness of the zone of interest then there is a certain probability that the well is not in the target because the well can be located anywhere within the cone. This is an example of how well placement adds significant value by steering the well according to the observed geology from the LWD responses rather than steering geometrically based on surveys with their associated uncertainties. As the reservoir targets being drilled become thinner and more complex, the ability to place wells based exclusively on the geometrical projection of surveys becomes increasingly risky.

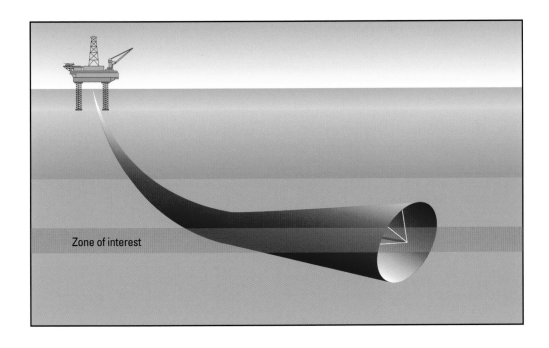

Zone of interest

Figure 4-17. The cone of uncertainty, shown in red, defines the volume within which the well is with a certain probability.

Figure 4-18. On the plan view of a geological target (square) superimposed on a reference coordinate grid, the planned well trajectory through the target (green line) has the ellipse of uncertainty (the horizontal projection of the ellipsoid of uncertainty) drawn around the planned well in the center of the target.

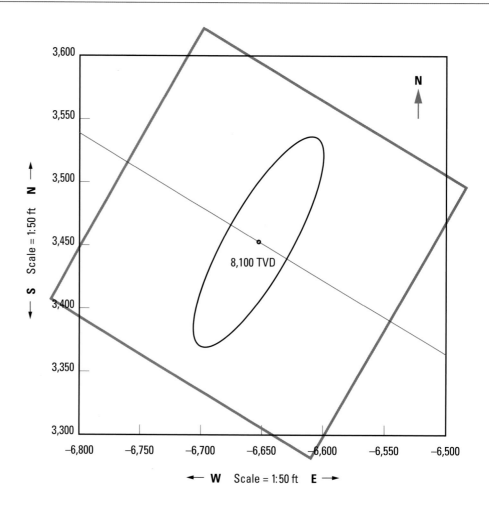

4.3.7 Geological and drilling targets

A geological target is usually specified based on the expected reservoir dimensions at a given location. The driller must assess whether the well can be drilled with sufficient accuracy to hit the target.

If the driller simply positioned the nominal location of the wellbore within the geological target, it is possible that part of the ellipse of uncertainty would fall outside of the geological target (Fig. 4-18). To ensure that the well is actually within the geological target the driller must reduce the target size by the wellbore ellipse of uncertainty, resulting in a smaller driller's target that ensures that the well, even if on the edge of the driller's target and on the edge of the ellipse of uncertainty, is within the geological target (Fig. 4-19).

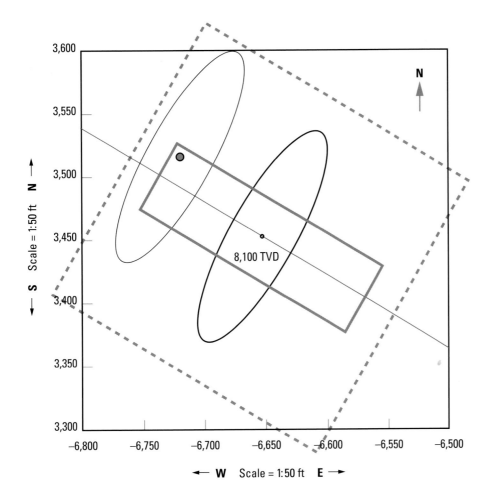

Figure 4-19. Shown in plan view, the driller's target (rectangle) within the geological target (dashed square) ensures that the well lands within the geological target.

4.3.8 Continuous inclination and azimuth

Static surveys are acquired while the drillstring is stationary to minimize possible noise on the measurements. It is possible to take both inclination and azimuth measurements while drilling; however, these continuous measurements are not as accurate as the static surveys and are not currently used for defining the well position. The two major advantages of continuous inclination and azimuth are that they allow users to observe changes in the well trajectory while drilling rather than having to wait for the next static survey and that they are sampled frequently, thereby overcoming the issue of wellbore tortuosity that is not captured by the relatively widely spaced static surveys (Fig. 4-20).

Figure 4-20. Continuous inclination and azimuth overcome the surveying phenomenon of wellbore tortuosity that is not captured by the widely spaced static surveys.

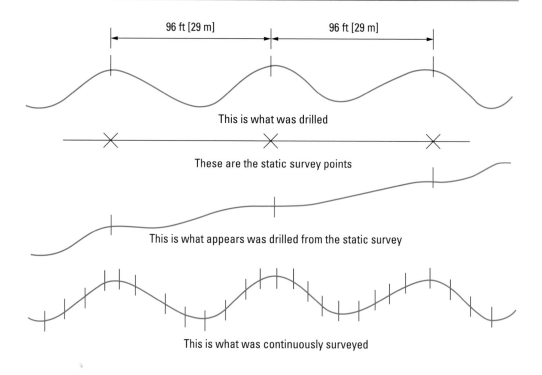

With the increasing use of rigs equipped with topdrive drilling systems, with which a 90-ft [27-m] stand of 3 joints of drillpipe can be drilled without having to add new drillpipe, the taking of a static survey every 30 ft [9 m] now appears to take rig time because adding a new joint of drillpipe each 30 ft is no longer required. To maximize ROP, drillers would prefer to drill the 90-ft stand between surveys.

The example in Fig. 4-21 demonstrates the inaccuracies that accumulate if the spacing between static surveys is allowed to extend to 90 ft. In the early part of the well the static surveys were taken every 30 ft, but under pressure to save time, the survey frequency dropped to 90 ft. The nature of the minimum curvature computation causes the computed position of the wellbore to appear deeper than that computed from the continuous survey data. The discrepancy increases with well depth, in this case accumulating a difference of over 5 ft [1.5 m] TVD after 2,500 ft [762 m] MD. In this situation, the continuous survey representation is likely to be more accurate.

An additional concern is that because the static survey is used in the geological model to define the 3D location of formation tops and fluid contacts, the TVD discrepancy could introduce errors into the static and dynamic models of the reservoir.

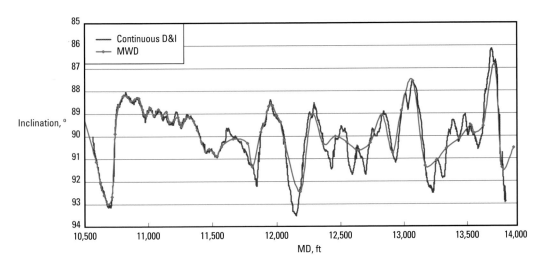

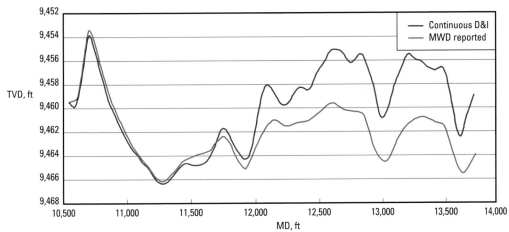

Figure 4-21. In this example of continuous inclination, the spacing of the static survey (red curve) in the upper panel was increased from about 30 ft to 90 ft from 11,600 ft [3,536 m] MD. The 90-ft spacing averages out many of the inclination changes, which are captured by the continuous inclination measurement (blue curve). The lower panel displaying the TVD computed from the static survey shows the well deeper than the depth computed from continuous inclination. The difference exceeds 5 ft TVD in places.

4.3.9 Projection of inclination to the bit

Because the survey package in the MWD tool is always located behind the motor or RSS in the BHA, the inclination of the bit must be calculated based on a projection of the drilling trend since the last static survey was taken. For example, if the driller has been holding inclination, then the bit inclination is expected to be the same as the last survey. If the driller has been building inclination at a rate of 1°/100 ft [30 m] for the 50 ft [15 m] since the last survey, then the inclination at the bit is projected to be 0.5° higher than the inclination at the last survey station. The drilling tendency is not always easy to predict, so the projected inclination at the bit is subject to uncertainty.

Inclination measurements taken with survey packages located close to the bit, such as those available with Schlumberger rotary steerable tools, significantly reduce the uncertainty on the bit inclination, enabling more accurate steering of the wellbore.

4.4 Downhole electrical power generation

Batteries could be used to deliver power to the measurement electronics and telemetry system, but the duration of drilling runs would be limited to the life of the batteries. For this reason many MWD systems incorporate a downhole mud turbine and alternator electrical power generation system. Whenever mud is being pumped through the drilling system, hydraulic power from the mud flow is converted into electrical power as the turbine rotates and drives the attached alternator (Fig. 4-22). This electrical power is then available to the MWD subsystems, and where an intertool electrical connection is available that provides power and data connectivity along the BHA, power from the MWD turbo-alternator system can also be used by other tools in the BHA.

Figure 4-22. Hydraulic power from the mud flow is converted into electrical power through use of a turbine and alternator system. The stator (blue) deflects the mud flow (from left to right) on to the rotor (red) causing it to turn. The alternator (not visible inside the yellow housing to the right) converts the rotation to electrical power.

4.5 Real-time telemetry

Real-time mud-pulse data telemetry techniques were originally developed to improve the efficiency of wellbore surveying while drilling, which had previously been acquired by the time-consuming process of running and retrieving single- or multishot mechanical surveying devices. The introduction of electronics and sensors capable of surviving the drilling environment, combined with the means to transmit the data to surface, significantly reduced the time required to survey a well. The original MWD application, then, was the acquisition and transmission of survey measurements.

There are now several methods for transmitting data from the downhole tools to surface:

- electromagnetic telemetry
- wired-drillpipe telemetry
- mud-pulse telemetry.

4.5.1 Electromagnetic telemetry

Electromagnetic telemetry uses voltage applied across an insulating gap on a drill collar to induce an electric field in the Earth which propagates to surface, where it is detected as a voltage difference between the drillstring and a receiver positioned away from the wellhead (Fig. 4-23). Data is encoded by modulating the signal frequency. The system is sensitive to the distribution of formation resistivity both around the tool and between the tool and surface. Before using electromagnetic telemetry, modeling of the system response in a given formation resistivity profile is required to determine if there will be sufficient signal detected at the surface. The major advantage of the system is that, unlike mud-pulse telemetry, electromagnetic telemetry does not depend on the presence of nearly incompressible flowing mud and so can be used when the mud pumps are off or when the mud is heavily gas-cut, as is often the case when drilling a well underbalanced (that is, the mud pressure is kept below the formation pressure so that formation fluid flows into the borehole while drilling).

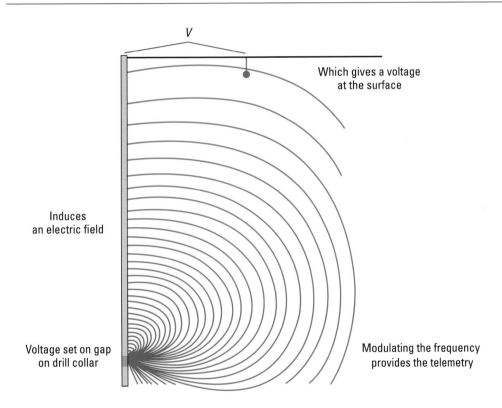

V

Which gives a voltage at the surface

Induces an electric field

Voltage set on gap on drill collar

Modulating the frequency provides the telemetry

Figure 4-23. Electromagnetic telemetry systems induce an electric field in the Earth which is detected at surface. Data is transmitted by modulating the frequency.

4.5.2 Wired-drillpipe telemetry

Wired-drillpipe telemetry is an emerging technology with the potential to significantly increase data transmission rates both from the tool to the surface and from the surface to the downhole tools. As the name suggests, wired-drillpipe telemetry uses drillpipe equipped with an insulated conductor running the length of the pipe. Inductive couplers at each end are screwed together to create the electrical connection between one joint of pipe and the next (Fig. 4-24). By creating a continuous electrical path between the surface and downhole tools, wired drillpipe promises to deliver high bidirectional telemetry rates independent of mud flow. The drillpipe connections must be extremely reliable because a single failure in any one of the several hundred joints of drillpipe breaks the connection and renders the system unusable.

Figure 4-24. Wired drillpipe uses inductive couplers at the pin (left) and box (right) of the drillpipe to make the electrical connection between one joint of drillpipe and the next.

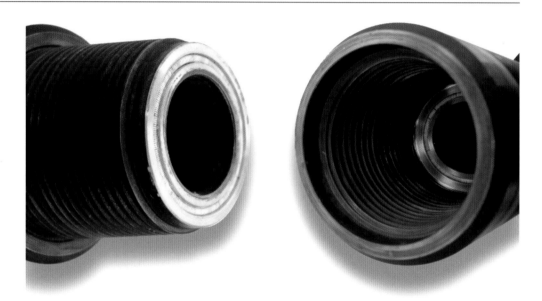

4.5.3 Mud-pulse telemetry

Mud-pulse telemetry is by far the most common real-time data transmission in use today. Mud-pulse telemetry involves encoding data in pressure pulses that propagate up through the mud inside the drillpipe. These pressure-pulse sequences are detected at the surface and decoded to recreate the numerical value of the data from the downhole tools. There are three main ways of creating a mud pressure pulse (Fig. 4-25). Positive-pulse systems impede mud flow with a poppet valve, resulting in a temporary increase in pressure. Negative-pulse systems use a bypass valve to bleed pressure off to the annulus, resulting in a temporary drop in pressure. Continuous wave carrier, or siren, systems use a rotating valve assembly that alternates between opened and closed positions, resulting in an oscillating pressure wave. Data can be encoded on the siren system by frequency or phase modulation.

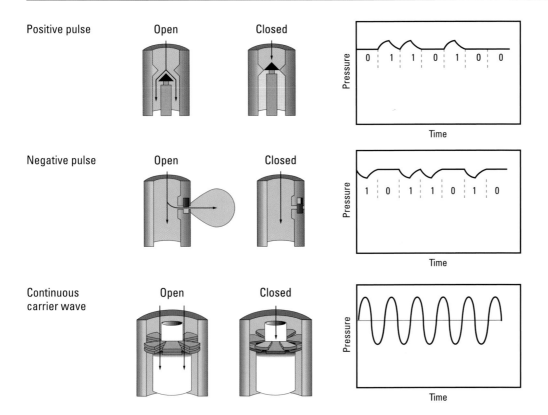

Positive pulse   Open   Closed

0 1 1 0 1 0 0

Negative pulse   Open   Closed

1 0 1 1 0 1 0

Continuous carrier wave   Open   Closed

Figure 4-25. The three main mud-pulse telemetry systems are positive pulse, negative pulse, and siren, which generates a continuous carrier wave.

Data acquired by the downhole LWD tools is compressed, encoded, and transmitted to the surface, most commonly through a mud-pulse telemetry system. The limited bandwidth of the current mud-pulse systems (typically 0.5 to 12 bits per second, or bps) limits the amount of data that can be transmitted to surface in real time. Improvements continue to be made in the telemetry rate (the number of bits that can be transmitted per second) and data compression (the amount of data transmitted per bit), but the selection of what data is sent is still required because bandwidth capable of transmitting all the data all the time, as is the case with wireline tools, is unlikely to be widely available in the near future.

4.5.3.1 Data points and frames

Real-time data is transmitted as data points (d-points), each of which represents a particular measurement (for example, formation bulk density) or is part of a larger collection of data such as part of an image spread across several d-points (Fig. 4-26).

The d-points are grouped into frames, which define the data to be transmitted when the BHA is in a particular mode of operation. For example, a frame designed for use when the BHA is sliding in a deviated well would contain the GTF and nonazimuthal formation measurements because there is no point wasting bandwidth by sending azimuthal data, which cannot be acquired when the BHA is not rotating. A frame designed for use when the BHA is rotating would, in contrast, most likely contain d-points for azimuthal and perhaps image data, but it would not contain a toolface d-point because the toolface is not of interest when the BHA is rotating.

Real-time data typically is transmitted in five designated frames:

- MTF frame—used in nearly vertical wells when sliding

- GTF frame—used in deviated and horizontal wells when sliding

- rotary frame—used when the BHA is rotating

Figure 4-26. The d-points containing the measurement data are grouped into frames.

Continous Survey and CDR* Compensated Dual Resistivity						
MTF		GTF		Rotary		Frame
25		26		27		
	bits		bits		bits	
SYNC/FID	13	SYNC/FID	13	SYNC/FID	13	Synchronization word and frame identifier
mtfs	6	gtfs	6	RP16H_c	6	
RP16H_c	8	RP16H_c	8	RA40L_c	8	
RA40L_c	8	RA40L_c	8	GRAPC_c	8	
GRAPC_c	7	GRAPC_c	7	RP16H_c	7	
mtfs	6	gtfs	6	RA40L_c	6	d-points
RP16H_c	8	RP16H_c	8	GRAPC_c	8	
RA40L_c	8	RA40L_c	8	rgx	8	
GRAPC_c	7	GRAPC_c	7	rhx	7	
mtfs	6	gtfs	6	RP16H_c	6	

- survey frame—used to transmit survey data acquired while the BHA was stationary and the mud pumps were off, which means that the mud-pulse telemetry was not operational. When the mud pumps are started at the surface and flow reestablished through the MWD tool, this is the first frame sent.

- utility frame—used to transmit additional data (other than the surveys) acquired while the mud pumps were off and hence the mud-pulse telemetry was not operational. This frame is used for applications such as hydrostatic mud pressure and transit times and waveforms from sonic and seismic measurements in addition to formation pressure data acquired with the pumps turned off to minimize measurement interference.

The selection of which frame to transmit is made by the downhole tool based on the well inclination (MTF or GTF frame), whether the tool is rotating (rotary frame), and whether the mud flow has just started again after a period of no flow (survey and utility frames). After a few training bits to allow the surface system to synchronize, the frame identifier is sent. Because both the surface and downhole systems were programmed with the same frames, the subsequent stream of bits is divided into the corresponding d-points and decoded by the surface system (Fig. 4-27).

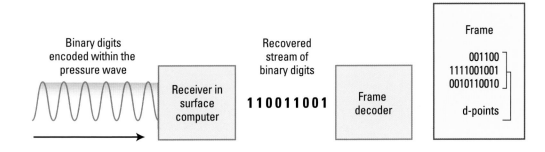

Figure 4-27. The bit stream encoded in the mud-pulse pressure wave is divided back into the d-points by the surface system, which was programmed with the same frame information as the downhole tool.

Decoding involves the conversion of the binary bit stream to decimal followed by application of the reverse of the transform that was applied to the data downhole. For example, a downhole density measurement of RHOB = 2.4 g/cm^3 could have 0.9 g/cm^3 subtracted and the remainder divided by 0.01 to give a decimal number of 150. The eight-bit binary equivalent, 10010110, is then transmitted via the mud-pulse system.

At the surface, the binary number 10010110 is converted back to the decimal number 150 and the reverse transform RHOB_RT = 0.01X + 0.9 g/cm^3 is applied (the "_RT" suffix identifies the measurement as real-time data to distinguish it from the recorded-mode RHOB measurement). The real-time measurement of RHOB_RT = 2.4 g/cm^3 is then available for visualization and interpretation. The names and descriptions of Schlumberger real-time channels are listed in the Schlumberger Curve Mnemonic Dictionary at http://www.slb.com/modules/mnemonics/.

4.5.3.2 Real-time and recorded-mode data

The data transmitted in real time by using mud-pulse telemetry is a small subset of the full suite of measurements recorded in the tool memory. The real-time data is the data transmitted to the surface shortly after being recorded along with the corresponding surface data (Fig. 4-28). The limited bandwidth of the mud-pulse telemetry reduces the number of channels, the sample update frequency, or both in the real-time data compared with the recorded-mode data. The recorded-mode data is retrieved from the memory of the downhole tools when the tools are brought back to the surface.

Figure 4-28. A limited amount of real-time data can be transmitted to the surface through the MWD telemetry. Recorded-mode data is retrieved from tool memory when the tools return to the surface and includes all the data from the full suite of measurements.

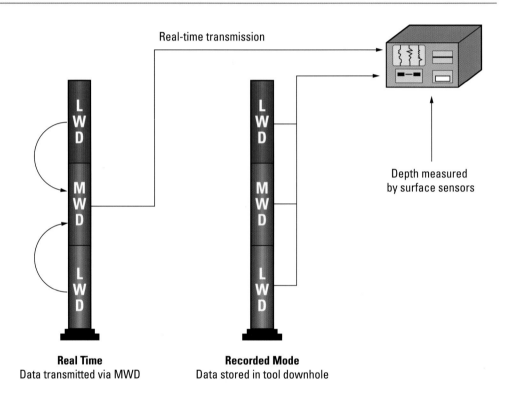

Real-time transmission

Depth measured by surface sensors

Real Time
Data transmitted via MWD

Recorded Mode
Data stored in tool downhole

4.5.3.3 Time-to-depth conversion

Downhole data is recorded versus time by using a downhole clock in the tool that is synchronized with a surface clock during programming of the tool. Depth is measured at the surface versus time, which is tracked with the surface synchronized clock. Real-time data is depth stamped on arrival at the surface and the data plotted versus either time or depth, depending on the measurement.

When the recorded-mode data is retrieved from the downhole tool, the measurements made at a particular time are merged with the recorded depth at the corresponding time to make the time-to-depth conversion of the data (Fig. 4-29). Because the measure points are distributed along the length of the LWD tools, their measurements are each at a different depth at a given time unless they are collocated, as is the case for the GR and resistivity measurements shown in Fig. 4-29.

This is an important concept for the quality control of log responses. Consider a GR sensor located 10 ft [3 m] from a density sensor on the BHA. The GR and density measurements are acquired at the same time but at different depths. If there is a problem with data transmission, the density and GR logs will be affected at different depths, indicating that the anomalous response is time related, not caused by a formation effect. If the GR and density both show a response at the same depth, then it is highly likely that it is due to a formation feature because the data was acquired at the same depth by the two measurements but at different times.

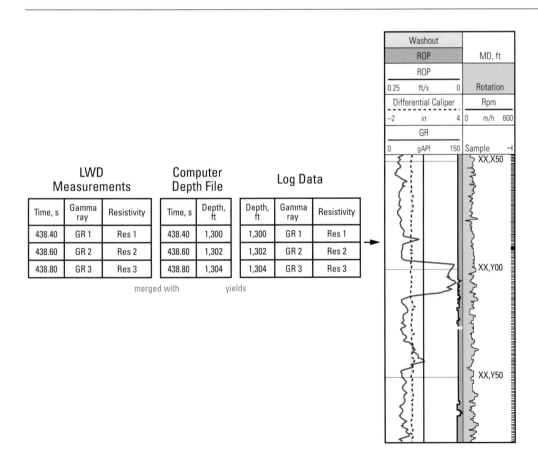

LWD Measurements

Time, s	Gamma ray	Resistivity
438.40	GR 1	Res 1
438.60	GR 2	Res 2
438.80	GR 3	Res 3

merged with

Computer Depth File

Time, s	Depth, ft
438.40	1,300
438.60	1,302
438.80	1,304

yields

Log Data

Depth, ft	Gamma ray	Resistivity
1,300	GR 1	Res 1
1,302	GR 2	Res 2
1,304	GR 3	Res 3

Figure 4-29. Data is recorded versus time downhole and depth is recorded versus time at the surface. The common time index is then used to merge the measurements with the corresponding depth to create the depth-indexed data for log display.

4.5.3.4 Time-lapse data

Because the downhole tools are continuously acquiring their measurements versus time, data is recorded to memory both while tripping in the hole and while pulling out. This means that there are at least two sets of measurements across a given depth. As the drillers make short wiper trips and ream it is possible that there are multiple passes over a given interval. This data, recorded over the same interval at different times, is called time-lapse data (Fig. 4-30). It enables tracking the evolution of formation changes as a function of time. Of particular interest are the changes associated with mud filtrate invasion of the formation because these can be diagnostic of the formation permeability. Also of interest are changes in the shape of the borehole or fracturing of the formation as it is subjected to different borehole pressures as the well is drilled deeper into the formation.

Because the formation is usually uninvaded during the drilling pass, the time-lapse data is a useful way to acquire an invaded-zone resistivity, R_{xo}, for use in evaluating what formation fluid was displaced by mud filtrate invasion and how much of the formation fluid is mobile.

Figure 4-30. At any given depth there are at least two and possibly more sets of measurements made at different times. Analyzing this time-lapse data enables the evaluation of changes in formation properties as a function of time, which may be caused by mud filtrate invasion or geomechanical changes to the borehole.

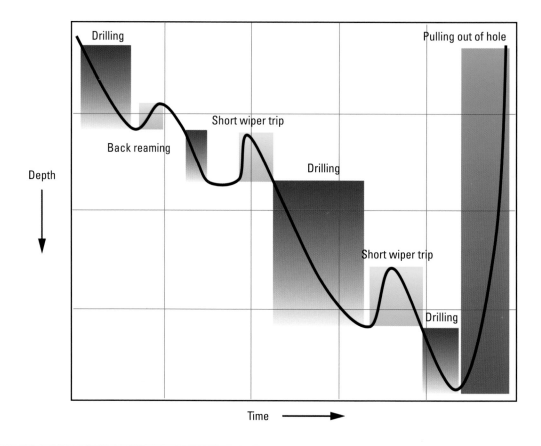

4.5.3.5 Real-time data rate

Each d-point requires a certain number of bits to encode the corresponding downhole measurement (for example, 7 bits for a standard GR d-point). The d-points of the measurements that are required in real-time are composed into frames with a certain size (236 bits for the examples shown in Fig. 4-31). The MWD system is configured to transmit a certain number of bits per second based on a combination of downhole and surface conditions, including the quality of the rig mud pumps, depth of the well, nature of the mud, and configuration of the MWD mud pulser. In the example in Fig. 4-31 the planned bit rate is 3 bps, so 236 bit frames take 78.67 s to transmit and each measurement gets one update each 78.67 s. Assuming that the average rate of penetration is 100 ft/h [30 m/h], this corresponds to 100 ft/3,600 s = 0.028 ft/s [0.008 m/s]. In 78.67 s the drillstring drills forward 2.06 ft [0.63 m], meaning that each sample is 2.06 ft apart, which corresponds to 0.48 samples/ft [1.57 samples/m]. This information is summarized in the "Update rate" columns in the lower panel of Fig. 4-31. If the required sampling rate is 1 sample/ft, then the telemetry rate must be increased to 6 bps, or the drilling slowed to 50 ft/h [15 m/h], or the number of d-point transmitted reduced so that the frame requires half the number of bits, or a combination of these measures.

In general, the real-time data rate can be calculated as

$$\frac{real\text{-}time\ samples}{distance\ \text{(ft)}} = \frac{3{,}600\ \text{(s/h)}}{ROP\ \text{(ft/h)}} \times \frac{bits\ per\ frame\ \text{(bits)}}{transmission\ rate\ \text{(bps)}} \times number\ of\ times\ d\text{-}point\ appears\ in\ the\ frame.$$

(4-5)

Figure 4-31. The real-time FrameBuilder module lists the MTF, GTF, and rotary (ROT) frames in the left panel. The right panel shows the corresponding time between measurement updates and the distance between data points for an assumed average drilling ROP of 100 ft/h.

FrameBuilder ImPulse* Frame Listing				Library:	
General Information		R.O.P.: 100.00 ft/h Bit Rate: 3.00 bps			Printed: 05/02/06 19:23:45
Frame No.: 200 Frame Type: MTF Frame Time: 78.67 s (236 bits)		Frame No.: 2001 Frame Type: GTF Frame Time: 78.67 s (236 bits)		Frame No.: 2002 Frame Type: ROT Frame Time: 78.67 s (236 bits)	
d-points List		**d-points List**		**d-points List**	
d-point	**Bits**	**d-point**	**Bits**	**d-point**	**Bits**
GRAPC_mx	7	GRAPC_mx	7	GRAPC_mx	7
RP16H_mx	8	RP16H_mx	8	RP16H_mx	8
RP22H_mx	8	RP22H_mx	8	RP22H_mx	8
RP40H_mx	8	RP40H_mx	8	RP40H_mx	8
RA22H_mx	8	RA22H_mx	8	RA22H_mx	8
RA40H_mx	8	RA40H_mx	8	RA40H_mx	8
ROBB_d	9	ROBB_d	9	ROBB_d	9
ROBU_d	9	ROBU_d	9	ROBU_d	9
TNRA_d	9	TNRA_d	9	TNRA_d	9
rgx	12	rgx	12	rgx	12
rhx	12	rhx	12	rhx	12
PERB_d	9	PERB_d	9	PERB_d	9
DRHB_d	8	DRHB_d	8	DRHB_d	8
DANG_mx	6	DANG_mx	6	DANG_mx	6
SPD4_mx	9	SPD4_mx	9	SPD4_mx	9
SAD4_mx	9	SAD4_mx	9	SAD4_mx	9
SPS4_mx	9	SPS4_mx	9	SPS4_mx	9
SAS4_mx	9	SAS4_mx	9	SAS4_mx	9
INCL_b	12	INCL_b	12	INCL_b	12
AZIMLO_b	10	AZIMLO_b	10	AZIMLO_b	10
STEER_b	8	STEER_b	8	STEER_b	8
TF_b	6	TF_b	6	TF_b	6
TFDS_b	6	TFDS_b	6	TFDS_b	6
PRDS_b	4	PRDS_b	4	PRDS_b	4
trpm	7	trpm	7	trpm	7
sticknslip	7	sticknslip	7	sticknslip	7
shkrsk	2	shkrsk	2	shkrsk	2
ADNSK_d	2	ADNSK_d	2	ADNSK_d	2
SHKLV_mx	2	SHKLV_mx	2	SHKLV_mx	2

Update Rates			Update Rates			Update Rates		
d-point	Interval, s	Spacing, ft	d-point	Interval, s	Spacing, ft	d-point	Interval, s	Spacing, ft
GRAPC_mx	74.33	2.06	GRAPC_mx	74.33	2.06	GRAPC_mx	74.33	2.06
RP16H_mx	74.33	2.06	RP16H_mx	74.33	2.06	RP16H_mx	74.33	2.06
RP22H_mx	74.33	2.06	RP22H_mx	74.33	2.06	RP22H_mx	74.33	2.06
RP40H_mx	74.33	2.06	RP40H_mx	74.33	2.06	RP40H_mx	74.33	2.06
RA22H_mx	74.33	2.06	RA22H_mx	74.33	2.06	RA22H_mx	74.33	2.06
RA40H_mx	74.33	2.06	RA40H_mx	74.33	2.06	RA40H_mx	74.33	2.06
ROBB_d	74.33	2.06	ROBB_d	74.33	2.06	ROBB_d	74.33	2.06
ROBU_d	74.33	2.06	ROBU_d	74.33	2.06	ROBU_d	74.33	2.06
TNRA_d	74.33	2.06	TNRA_d	74.33	2.06	TNRA_d	74.33	2.06
rgx	74.33	2.06	rgx	74.33	2.06	rgx	74.33	2.06
rhx	74.33	2.06	rhx	74.33	2.06	rhx	74.33	2.06
PERB_d	74.33	2.06	PERB_d	74.33	2.06	PERB_d	74.33	2.06
DRHB_d	74.33	2.06	DRHB_d	74.33	2.06	DRHB_d	74.33	2.06
DANG_mx	74.33	2.06	DANG_mx	74.33	2.06	DANG_mx	74.33	2.06
SPD4_mx	74.33	2.06	SPD4_mx	74.33	2.06	SPD4_mx	74.33	2.06
SAD4_mx	74.33	2.06	SAD4_mx	74.33	2.06	SAD4_mx	74.33	2.06
SPS4_mx	74.33	2.06	SPS4_mx	74.33	2.06	SPS4_mx	74.33	2.06
SAS4_mx	74.33	2.06	SAS4_mx	74.33	2.06	SAS4_mx	74.33	2.06
INCL_b	74.33	2.06	INCL_b	74.33	2.06	INCL_b	74.33	2.06
AZIMLO_b	74.33	2.06	AZIMLO_b	74.33	2.06	AZIMLO_b	74.33	2.06
STEER_b	74.33	2.06	STEER_b	74.33	2.06	STEER_b	74.33	2.06
TF_b	74.33	2.06	TF_b	74.33	2.06	TF_b	74.33	2.06
TFDS_b	74.33	2.06	TFDS_b	74.33	2.06	TFDS_b	74.33	2.06
PRDS_b	74.33	2.06	PRDS_b	74.33	2.06	PRDS_b	74.33	2.06
trpm	74.33	2.06	trpm	74.33	2.06	trpm	74.33	2.06
sticknslip	74.33	2.06	sticknslip	74.33	2.06	sticknslip	74.33	2.06
shkrsk	74.33	2.06	shkrsk	74.33	2.06	shkrsk	74.33	2.06
ADNSK_d	74.33	2.06	ADNSK_d	74.33	2.06	ADNSK_d	74.33	2.06
SHKLV_mx	74.33	2.06	SHKLV_mx	74.33	2.06	SHKLV_mx	74.33	2.06

The higher the telemetry rate, the more data can be transferred, the faster drilling can proceed, and the higher the rate at which data can be sampled. Increased telemetry rates can be achieved by increasing the physical number of mud-pulse bits per second or by using data compression techniques to increase the amount of information encoded in each physical mud-pulse bit. Data compression results in a higher effective transmission rate without changing the physical telemetry rate (Fig. 4-32).

The increasing use of data compression now enables the real-time transmission of images, waveforms, and array data, which greatly enhances the capabilities of real-time interpretation.

Figure 4-32. Increasing the telemetry rate by a factor of 4 through data compression is the equivalent of increasing the physical bit rate by a factor of 4, either of which allows transmission of 4 times the amount of data.

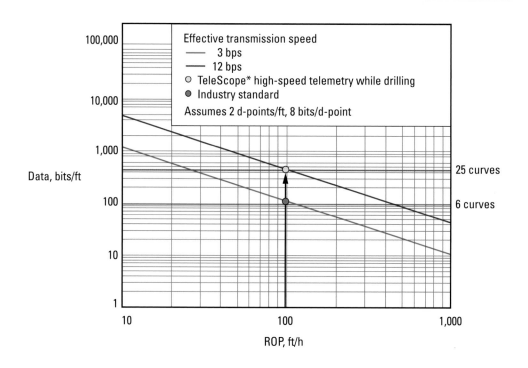

4.5.3.6 Recorded-mode data rate

The recorded-mode data rate follows a similar logic to the real-time data rate. The main difference is that the update rate is not defined by the size of the frame and the real-time data transmission rate. In recorded mode the update rate is defined by how frequently the downhole tool takes the measurement of interest. For example, a density measurement could be acquired over a period of 10 s or 20 s. Obviously there are twice as many samples of the 10-s density measurement compared with the 20-s measurement (although the statistical precision of the 10-s measurement is lower). For an average ROP of 100 ft/h, this corresponds to 100 ft/3,600 s = 0.028 ft/s, or 0.28 ft per sample for the 10-s measurement and 0.56 ft per sample for the 20-s measurement.

Another way to express this is that there are 3.6 samples per foot for the 10-s measurement and 1.8 samples per foot for the 20-s measurement.

In general, the recorded-mode data rate can be calculated as

$$\frac{recorded\text{-}mode\ samples}{distance\ \text{(ft)}} = \frac{3{,}600\ \text{(s/h)}}{ROP\ \text{(ft/h)}} \times \frac{1}{update\ rate\ \text{(s)}}. \tag{4-6}$$

4.6 Drilling-related measurements

Numerous measurements are taken at the surface during drilling to aid the driller in understanding what is happening downhole:

- depth—length of drillpipe run in the hole

- ROP—how far has been drilled per unit time

- mud flow rate—into and out of the hole, with the difference indicating mud loss to the formation

- surface torque—applied to the drillpipe

- surface weight—on the traveling block, which holds the drillstring

- standpipe pressure—the mud pressure as it enters the drillpipe at the surface

- surface drillpipe rpm.

In many cases there are complementary measurements made downhole by the MWD tool that assist in clarifying the behavior of the downhole system (Fig. 4-33).

Figure 4-33. A drilling performance log presents downhole measurements and where possible compares them with their surface equivalents.

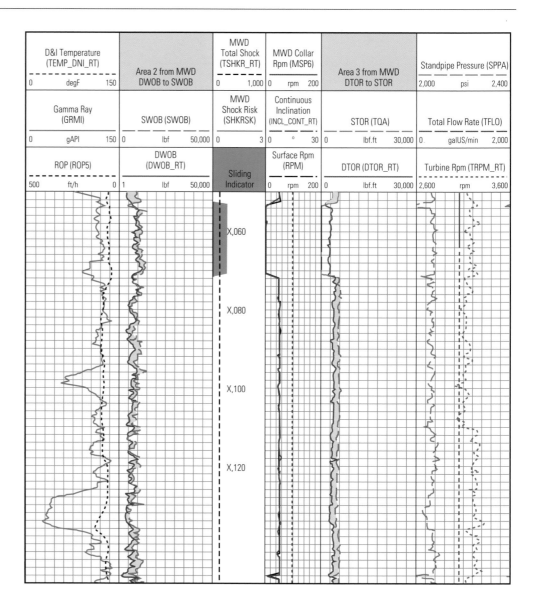

4.6.1 Downhole temperature

The downhole temperature is measured in several places along the toolstring. The D&I temperature is the internal temperature of the accelerometer and magnetometer package. It is used to correct for temperature-related drift in the sensor responses. It is also used to ensure that the tool is being run within its temperature rating. The D&I temperature is displayed in the left track of Fig. 4-33.

The temperature of the mud in the annulus is measured at the same point as the annular pressure. The temperature is used to correct the pressure gauge for temperature effects. The mud temperature measured in the annulus is usually lower than the internal tool temperature owing to heat dissipation from the electronics in the tool. The mud temperature can be significantly lower than the static formation temperature because the mud, which is relatively cool at the surface, is pumped through the drillstring faster than it can reach thermal equilibrium with the surrounding formation. Hence, the mud has a cooling effect on the formation.

4.6.2 Surface and downhole weight on bit

As drillpipe is run in the hole, part of the weight of the drillpipe is taken up by the buoyancy effect as it displaces mud in the hole. As more drillpipe is run in the hole, the weight on the traveling block, which holds the drillstring, increases to hold the buoyant weight of the drillstring. Just before the bottom of the hole is touched, the weight on the traveling block is at a maximum. When the bit rests on the bottom of the hole, the weight decreases, indicating that some of the string weight is now being taken by the bit. The difference between the buoyant weight of the drillstring in the hole and the measured weight on the traveling block indicates the weight applied by the bit to the formation plus any weight that is being held by friction along the borehole wall. The difference between surface weight off bottom and on bottom is called the surface weight on bit (SWOB). This is a key drilling parameter that can be controlled by the driller at the surface by pulling up on the traveling block to reduce weight on bit or slacking off to increase weight on bit.

Friction between the drillstring and the borehole can take up some of the weight reduction seen at the surface. By measuring the actual weight applied to the bit downhole, the driller determines how much weight is being lost to friction.

The second track in Fig. 4-33 compares the downhole weight on bit (DWOB, blue solid curve) and the computed SWOB (red dashed curve). The shading between them highlights any increase in separation, which indicates an increase in friction along the borehole.

Poor weight transfer to the bit may be due to cuttings buildup in the hole, BHA diameter changes (for example, stabilizers) hanging on ledges in the borehole, borehole collapse, or sharp doglegs, which a relatively stiff BHA may have difficulty bending through. Particularly in high-angle and horizontal holes, frictional effects can become so great that the BHA cannot move forward. This is called lock-up (Fig. 4-34). Any further release of weight from the surface results in drillpipe buckling, meaning that the well cannot be drilled any farther.

Frictional effects are greater when the drillstring is sliding than when it is rotating. This is because rotation keeps a film of mud between the drillstring and the borehole wall which lubricates any motion. If the drillstring is static, then there is no film of mud, so the static coefficient of friction is higher than the dynamic coefficient of friction. This is one of the reasons that RSSs with their continuous rotation are able to drill longer, more complicated horizontal wells than mud motors with their slide-rotate sequences.

Figure 4-34. Lock-up occurs when frictional forces prevent the further transfer of weight to the bit. Any subsequent release of weight from the surface results in the drillpipe buckling.

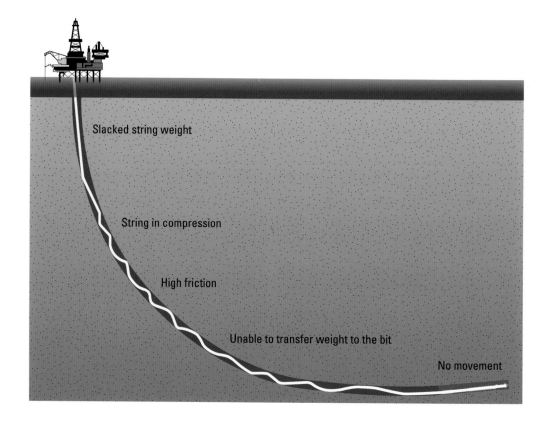

Slacked string weight

String in compression

High friction

Unable to transfer weight to the bit

No movement

4.6.3 Surface and downhole rpm

If the drillpipe is being rotated from the surface (which is not the case during sliding with a mud motor), then the surface rpm can be measured. Orthogonal magnetometers in the downhole tools measure sinusoids as the tool rotates in the Earth's magnetic field (Fig. 4-35).

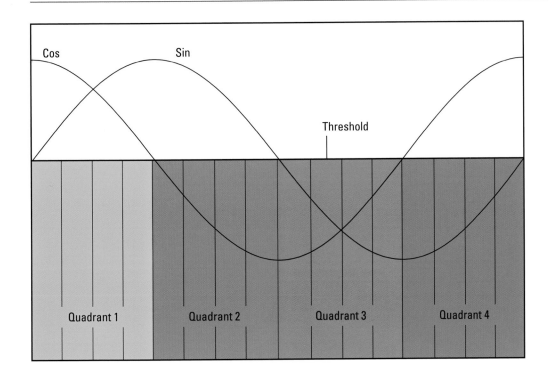

Figure 4-35. Orthogonal magnetometers measure 90°-shifted sinusoidal responses as they are rotated in the Earth's magnetic field, which can be used to determine the number of rpm of the drill collar. This type of magnetometer information is also used to determine the orientation of the LWD collar when azimuthal measurements are made.

While the average number of rpm at the surface and downhole must match, frictional effects can cause the drillpipe to slow and even stop temporarily while turns continue to be put in at surface. Eventually the torque overcomes the friction and the downhole assembly spins rapidly as the torque stored in the drillpipe is released. This stick-slip phenomenon reduces the efficiency of drilling and can cause severe damage to downhole tools and the bit.

In Fig. 4-33 the surface and downhole rpm are presented in the fourth track, overlaid on the same scale. The downhole rpm curve is smooth, indicating smooth rotation of the BHA. Erratic or "noisy" downhole rpm is diagnostic of stick-slip.

Rpm is also a useful sliding indicator. The rpm dropping to zero but hole depth increasing indicates that the BHA is sliding. An example of this is shown from X,053 to X,071 ft in Fig. 4-33.

4.6.4 Torque

Torque can be thought of as "rotational force," or "angular force," that causes a change in rotational motion. Torque, τ, is given by linear force multiplied by a radius (Fig. 4-36):

$$\tau = F \times r, \tag{4-7}$$

where

F = component of the force vector applied perpendicular to the moment arm
r = moment arm along which the torque is applied.

The SI unit for torque is the newton-meter (N.m). In customary oilfield units, torque is measured in the pound-foot (lbf.ft) (also known as the foot-pound).

In a drilling environment torque is applied to rotate drillpipe and the bit. Surface torque applied to rotate the drillstring can be measured and plotted alongside the downhole torque measured near the bit. An example of this is shown in the second track from the right of Fig. 4-33, where the surface torque (STOR, red dashed curve) and downhole torque (DTOR, blue solid curve) are presented on the same scale. The small separation between them is due to frictional losses of torque against the wall of the borehole and around doglegs in the well trajectory. A significant increase in the separation indicates that the surface torque is not being transferred smoothly to the bit, possibly because of a worn bit or cuttings buildup. In extreme cases, this frictional torque can prevent rotation of the drillpipe.

In the sliding section from X,053 to X,071 ft on Fig. 4-33, DTOR does not drop to zero when the surface torque is no longer applied. This is because the mud motor is supplying the torque downhole to rotate the bit. Hence, DTOR is present even though no surface torque is being applied.

Figure 4-36. Torque is the product of force and length.

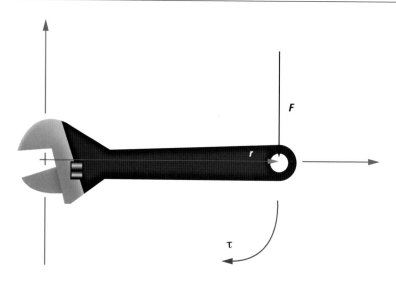

4.6.5 Surface and downhole mud pressure

Pressure drives the mud flow from the mud pumps, through the drillpipe and BHA, out through the bit, and back up the annulus between the drillpipe and borehole wall to the surface (Fig. 4-37). The mud pressure is measured in the standpipe that carries the mud up the side of the rig derrick from where it enters the drillstring. The standpipe pressure (SPP) can be thought of as the output pressure from the mud pumps.

The mud pressure increases as the mud travels down inside the drillpipe according to the equation

$$internal\ hydrostatic\ pressure = MW_{IN} \times g_n \times TVD, \qquad (4\text{-}8)$$

where

MW_{IN} = mud weight (density) as the mud is pumped into the hole

g_n = acceleration due to gravity

TVD = true vertical depth of the point where the pressure is being determined.

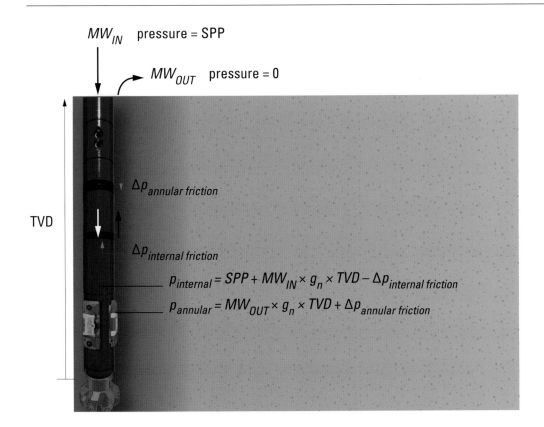

MW_{IN} pressure = SPP

MW_{OUT} pressure = 0

TVD

$\Delta p_{annular\ friction}$

$\Delta p_{internal\ friction}$

$$p_{internal} = SPP + MW_{IN} \times g_n \times TVD - \Delta p_{internal\ friction}$$

$$p_{annular} = MW_{OUT} \times g_n \times TVD + \Delta p_{annular\ friction}$$

Figure 4-37. Annular and internal mud pressure drive the mud flow.

When the mud is flowing, the pressure is increased by the standpipe pressure applied at the surface and decreased by frictional pressure losses along the inside of the drillstring:

internal circulating pressure = SPP + internal hydrostatic pressure − internal frictional losses. (4-9)

Additional pressure losses occur across the BHA, where components such as the MWD mud turbine and the drilling mud motor extract power from the mud flow. There is also a significant pressure loss across the nozzles in the bit, which accelerate the mud flow to clean the bit and agitate the rock cuttings into the mud flow for transport back to the surface.

A strain gauge measures the mud pressure inside the tool, before the pressure loss across the mud motor and bit occurs. As the mud flows back up the annulus between the drillstring and the borehole wall, another strain gauge measures the annular mud pressure.

The difference between the internal and annular pressures is primarily due to the pressure loss across the mud motor or RSS and the bit. Changes in the pressure difference can help in diagnosing performance issues with the mud motor, RSS, and bit nozzles.

Mud returning to surface is at atmospheric, or zero gauge, pressure as it flows into the mud tanks. The annular hydrostatic pressure at a point in the well is given by

$$\textit{annular hydrostatic pressure} = MW_{OUT} \times g_n \times TVD, \tag{4-10}$$

where

MW_{OUT} = mud weight (density) of the mud in the annulus. This is greater than MW_{IN} as a result of the rock cuttings that are suspended in the mud returning to the surface.

To overcome the annular frictional forces acting against the flow, the pressure at a given depth must be increased so that the mud arrives at the surface with zero pressure:

annular circulating pressure = annular hydrostatic pressure + annular frictional losses. (4-11)

The equivalent circulating density (ECD) of the mud in the annulus includes the effect of cuttings in circulation and frictional effects:

$$ECD = \textit{annular circulating pressure}/(g_n \times TVD). \tag{4-12}$$

When the mud pumps are off and there is no mud flow, the frictional forces drop to zero and some of the cuttings fall out of suspension, although some of the smaller cuttings remain in suspension. The equivalent static density (ESD) of the mud in the annulus is given by

$$ESD = \textit{annular hydrostatic pressure}/(g_n \times TVD). \tag{4-13}$$

The driller watches the ECD to ensure that the hole is being cleaned effectively and that cuttings are not building up against the BHA downhole, a situation that could lead to stuck pipe. Monitoring the ECD also allows timely adjustments to the mud density to maintain pressure control of the borehole.

Wellbore pressure control is critical for safe and smooth drilling operations. If the mud pressure is too high the formation may fracture, resulting in the loss of borehole fluid and subsequent drilling problems. If the mud pressure is too low then the fluid in the formation may flow into the well and begin migrating to surface, resulting in a "kick" or loss of pressure control. Depending on the mechanical properties of the rock, too high or too low a mud weight may result in a variety of borehole failure modes, most of which are detrimental to efficient drilling.

Well pressure control design involves determining the upper and lower pressure limits within which the well can be drilled safely. These limits are usually expressed as fluid densities. The upper limit is called the fracture gradient and defines the fluid density above which the formation fractures, resulting in mud loss. The lower gradient is usually the pore pressure gradient, below which formation fluid flows into the borehole. During drilling, the ECD must be kept within the safe drilling window (Fig. 4-38).

The overburden pressure (purple curve) is the fracture gradient and defines the upper limit of the pressure window in Fig. 4-38. The predrill seismic estimate of the pore pressure (black curve) defines the pressure window's lower limit. The closeness of the two curves indicates that there is little margin for error in mud weight. The resistivity-derived pore pressure is shown in red and the actual mud weight profile, plotted as the annular pressure-derived ECD, is in blue. Overall, the drilling plan succeeded in staying within the narrow pressure window. However, at two depths where the mud weight dropped below the lower pressure limit, the well took kicks.

4.6.6 Mud flow rate and turbine rpm

Within the flow range of the MWD tool configuration, the turbine rpm (TRPM) is linearly related to the mud flow rate through the tool. For a given mud flow rate into the drillpipe at the surface, the TRPM can be determined and monitored during real-time operations. If the TRPM decreases without a corresponding reduction of the flow rate into the drillpipe at the surface, this can be an early indicator that a hole (called a washout) has developed in the drillstring somewhere between surface and the MWD tool. This is vital information for the driller because leaks in the drillpipe can quickly erode into significant holes that weaken the drillpipe and may lead to the drillstring parting, leaving the BHA in the hole and resulting in expensive fishing operations.

The right track of Fig. 4-33 displays the total flow rate (blue dashed curve), SPP (red dashed curve), and TRPM (green dotted curve). The TRPM follows the SPP whereas the total flow is heavily averaged.

Figure 4-38. Where the mud weight dropped below the lower pressure limit of the pressure window, the deepwater well took kicks.

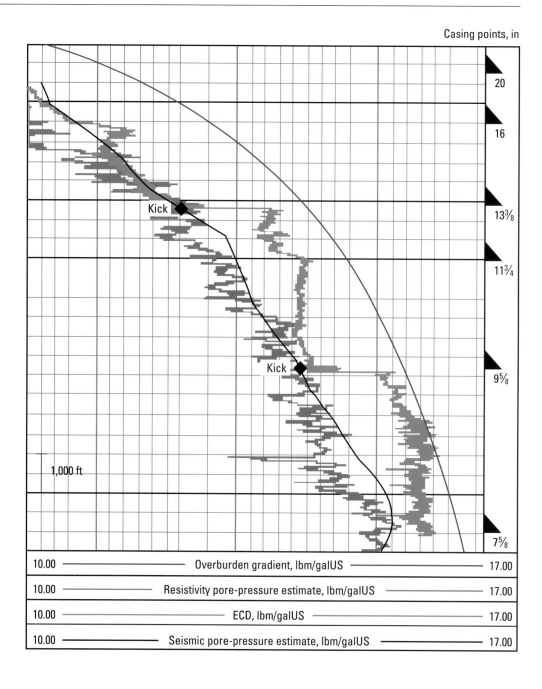

Casing points, in

	20
	16
Kick	13⅜
	11¾
Kick	9⅝
1,000 ft	
	7⅝

10.00	Overburden gradient, lbm/galUS		17.00
10.00	Resistivity pore-pressure estimate, lbm/galUS		17.00
10.00	ECD, lbm/galUS		17.00
10.00	Seismic pore-pressure estimate, lbm/galUS		17.00

4.6.7 Shock and vibration

Shock occurs in a drilling environment as the sudden input of energy when the BHA, bit, or drillstring impacts the borehole. A shock peak and the number of shocks per second over a specified threshold are the shock parameters normally measured (Fig. 4-39). Modern tools measure the shock in three dimensions so that the orientation of the impact that caused the shock is known.

Vibration can be thought of as the cumulative energy in the shock or root mean square (rms) of the drillstring response to the shock.

Shock and vibration can cause failure or damage to the BHA (collars, stabilizers, connections, and downhole tools) and drilling bit. The potential cost impact when components in the BHA are affected by shock and vibration is huge. Examples of these costs include

- extra rig cost from tripping a failed BHA out of the hole and running in with a new one

- twist-off connections, leading to a fishing operation or lost-in-hole charge for the BHA

- overgauge hole, which increases mud and cement volumes

- inability to evaluate the reservoir as a result of poor hole quality and severely degraded formation evaluation measurements.

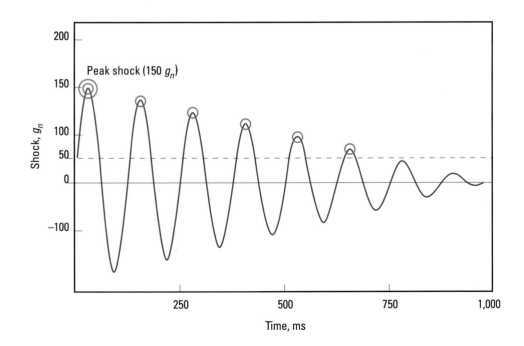

Figure 4-39. Shock is generally quantified by peak shock (measured in g_n) and the number of shocks measured over a specified threshold.

Shock and vibration measurements can be used to detect both good and bad drilling practices and trends that could lead to problems. The goal is to allow drilling to continue in the safest, efficient manner. Detection, understanding, and mitigation of shock and vibration are a way to achieve this goal.

Monitoring drilling mechanics, diagnosis of the cause of unwanted BHA behavior, and subsequent mitigation of shock and vibration improve the ROP through ensuring that energy is transmitted smoothly to the cutting surfaces of the bit rather than being wasted in damaging the BHA (Figs. 4-40 and 4-41). In turn, the total time and cost to drill the well are reduced.

Figure 4-40. A normal amount of shock and vibration occurs during the drilling process as energy is used mainly to drill rock and hence maximize the ROP.

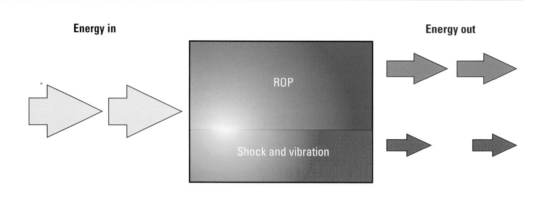

Figure 4-41. Shock and vibration in excess of what normally occurs takes energy away from the drilling process, thereby reducing the ROP.

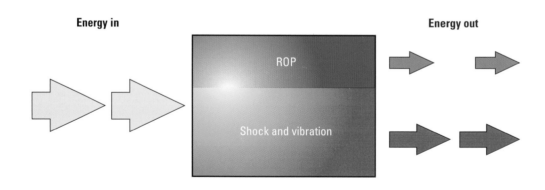

Shock and vibration are generally caused by BHA resonances resulting from interactions with the borehole. These can be axial, such as bit bounce; torsional, such as stick-slip; or lateral, such as forward and backward whirl (Fig. 4-42). These complex interactions between the rotating drill collars and the borehole wall vary with the weight on bit, rpm, lubricity of the mud, and flexibility of the BHA, among other variables. Diagnosis of the exact cause of shock can be difficult; however, having downhole measurements to warn of potentially destructive BHA behavior alerts the driller that remedial action must be taken.

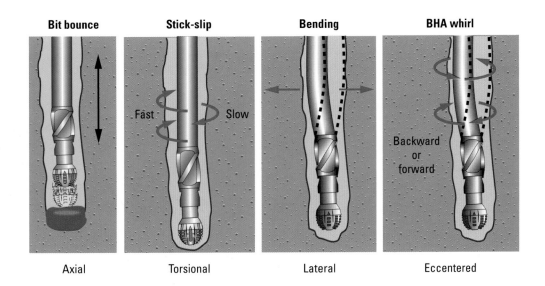

Figure 4-42. Various BHA resonances can cause shocks and vibrations.

Logging-While-Drilling
Fundamentals

Logging while drilling *is the measurement of formation properties during the deepening of the borehole or shortly thereafter through the use of measurement tools integrated into the BHA.*

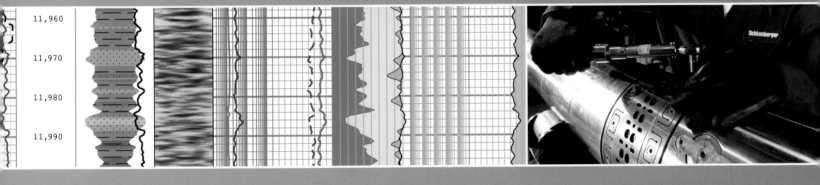

11,960

11,970

11,980

11,990

5.1 Definition

Logging while drilling is the measurement of formation properties during the deepening of the borehole or shortly thereafter through the use of measurement tools integrated into the BHA.

5.2 LWD acquisition considerations

In comparison with wireline acquisition, LWD has the advantages of measuring formation properties before drilling fluids invade deeply and being acquired during the drilling process so that it requires only minimal additional rig time. Further, many wellbores—especially highly deviated wells—prove to be difficult or even impossible to measure with conventional wireline tools. In these situations, LWD acquisition ensures that measurements of the subsurface are captured in the event that wireline operations are not possible.

As LWD measurements are made in the dynamic drilling environment they are subject to the effects of irregular BHA motion such as shock and stick-slip. Changing mud conditions such as mud temperature, density, and resistivity variations during drilling may also affect LWD logs. Generally, borehole conditions are more stable during wireline acquisition.

Unlike wireline logging, where a standard tool size is run in a wide range of borehole diameters, LWD tools must have the same diameter as the other collars in the BHA, which vary from one hole diameter to the next. Consequently, LWD tools come in a range of sizes. The diameter in inches of the LWD collar is typically specified as part of the name of the tool. For example, the 4¾-in- [12.1-cm-] diameter arcVISION Array Resistivity Compensated tool is named the arcVISION475* tool.

Table 5-1 shows the common tool sizes and normal borehole diameter range in which they are operated.

Table 5-1. Common LWD Tool Sizes and Corresponding Hole Diameters

Nominal Tool Diameter (Mnemonic), in [cm]	Common Hole Diameter Range, in [cm]
3⅛ (312) [7.9]	3¾ to 5⅞ [9.5 to 14.9]
4¾ (475) [12.1]	5¾ to 6¾ [14.6 to 17.1]
6¾ (675) [17.1]	8¼ to 9⅞ [20.9 to 25.1]
8¼ (825) [20.9]	10½ to 12¼ [26.7 to 31.1]
9 (900) [22.9]	12¼ to 17½ [31.1 to 44.4]

5.3 Petrophysics fundamentals

The conventional objective of logging, on wireline or while drilling, is to evaluate the volume, properties, and producibility of any hydrocarbons in the formation. With the introduction of LWD measurements, real-time formation evaluation for the purpose of optimally placing the well in the reservoir has become possible.

Since it was first published[†] in 1942, Archie's equation and its variants have been the standard method for evaluating formation water and hence hydrocarbon saturation. Gus Archie proposed the following basic form of the equation based on empirical correlation to experimental data:

$$S_w = \sqrt[n]{\frac{a}{\phi^m} \frac{R_w}{R_t}},$$

(5-1)

where

S_w = formation water saturation
a = empirically derived constant
m = cementation exponent
n = saturation exponent
ϕ = formation porosity
R_w = in situ water resistivity
R_t = uninvaded formation resistivity.

The classic triple-combo measurements (resistivity, density, and neutron) are the minimum set of inputs to solve for water saturation. Usually the bulk density measurement, ρ_b, is used to derive the formation porosity:

$$\rho_b = (\phi \times \rho_{fluid}) + (1 - \phi)\rho_{matrix}$$

(5-2)

$$\Longrightarrow \quad \phi_d = (\rho_{matrix} - \rho_b)/(\rho_{matrix} - \rho_{fluid}),$$

(5-3)

where

ϕ_d = density porosity
ρ_b = bulk density
ρ_{fluid} = fluid density
ρ_{matrix} = matrix density.

The neutron and density measurements are usually presented on a lithology-compatible scale (for example, limestone-compatible scale) such that in a freshwater-filled formation of the selected lithology, the neutron and density measurements overlie. Any separation between the two curves is an indication that either the fluid in the pore space is not fresh water or the matrix is not that assumed to create the overlay scale. These separations help identify when the ρ_{fluid} or ρ_{matrix} terms in the density-porosity Eq. 5-3 need to be reviewed to obtain the correct porosity.

[†] Archie, G.E.: "The Electrical Resistivity Log as an Aid in Determining Some Reservoir Characteristics," *Transactions AIME* (1942) 146, 54–61.

The photoelectric factor (PEF) and its volumetric equivalent, U, which are generally available with conventional density measurement tools, add information to help evaluate complex lithologies so that the correct matrix density can be entered into the density-porosity equation and hence an accurate porosity determined.

The resistivity measurements are used to evaluate the true formation resistivity, R_t. In consideration of the invasion of mud filtrate into the formation during drilling, modern resistivity tools measure the formation resistivity at multiple depths of investigation to characterize and correct for the near-wellbore invasion (Fig. 5-1). For LWD measurements close to the bit the invasion is often minimal, allowing direct formation measurements and revealing differences with subsequent wireline measurements.

In addition to the formation porosity, determined using a combination of the density, neutron, and photoelectric measurements, and a formation resistivity representative of the uninvaded formation resistivity, the Archie parameters a, m, and n need to be known (for example, from core analysis) or assumed (the default values are 1, 2, and 2, respectively). The formation water resistivity, R_w, at downhole conditions also needs to be known. R_w is usually calculated based on downhole temperature and pressure and the water salinity determined from water samples.

Archie's equation can then be solved to find the proportion of the pore space filled with water, otherwise known as the water saturation, S_w. The remaining pore space is assumed to be filled with hydrocarbons. Hence, the hydrocarbon saturation, S_{hc}, is determined as

$$S_{hc} = 1 - S_w. \tag{5-4}$$

The volume of hydrocarbons per unit volume of formation, V_{hc}, is given by the porosity of the formation multiplied by the hydrocarbon saturation:

$$V_{hc} = \phi S_h. \tag{5-5}$$

Given the volume of the reservoir, usually derived from seismic interpretation along with well-to-well log correlation, the total volume of hydrocarbons in place is the volume of the reservoir multiplied by the volume of hydrocarbons per unit volume of formation. This hydrocarbon volume is generally corrected for the change in volume of the hydrocarbons as they move from downhole to surface conditions. If the hydrocarbon is oil, the volume is given as the stock-tank oil in place (STOIP). If it is gas, the volume is quoted in standard cubic feet (scf) or standard cubic meters (scm):

$$STOIP \text{ or } scf = V_{hc} \times V_{reservoir} \times B, \tag{5-6}$$

where

$V_{reservoir}$ = reservoir volume

B = coefficient accounting for the change in volume that occurs when the hydrocarbons move from downhole to surface conditions.

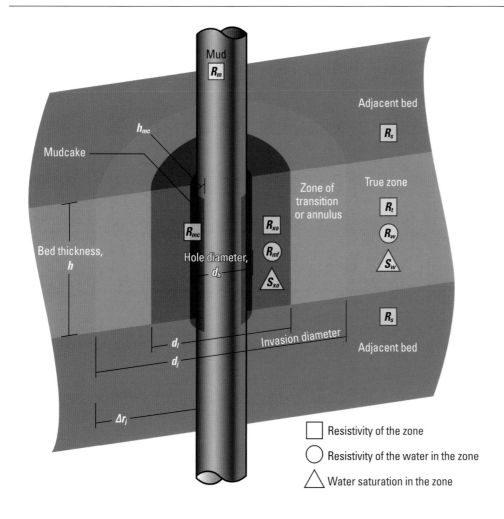

Mud
R_m

Adjacent bed
R_s

h_{mc}

Mudcake

Zone of transition or annulus

True zone
R_t
R_w
S_w

R_{mc}

R_{xo}

Bed thickness, h

Hole diameter, d_h

R_{mf}

S_{xo}

R_s

d_I

Invasion diameter

d_j

Adjacent bed

Δr_i

☐ Resistivity of the zone

◯ Resistivity of the water in the zone

△ Water saturation in the zone

Figure 5-1. Mud filtrate invasion creates an altered zone near the wellbore for which the measurements must be corrected. LWD measurements generally do not require as much correction as later wireline measurements because the invasion is not as deep soon after the hole has been drilled.

The porosity and saturation derived from triple-combo data are crucial in evaluating whether hydrocarbons are present and if so what volume of hydrocarbons and thus have a critical role in the decision to develop a reservoir or not.

As outlined previously, the original objective of migrating logging measurements from wireline tools to drill collars was to obtain the inputs for formation saturation evaluation with minimal additional rig time, less invasion, and better borehole conditions than are generally present when data is acquired after the formation has been open to fluid invasion and borehole degradation.

With the introduction of MWD and LWD tools in the late 1980s it became apparent that the application of LWD measurements could take advantage of the unique real-time capabilities. LWD measurement utilization expanded from formation evaluation to real-time structural assessment of the well position within the geological sequence and subsequent adjustment of the drilling trajectory to place the well in the desired location.

While the formation measurements are made downhole, depth and a number of measurements derived from it are measured at surface. Unlike wireline acquisition, where high-telemetry bandwidth enables all the data to be transmitted and depth stamped almost instantaneously, LWD measurements are recorded downhole against time while depth is recorded at surface against time and the time index is used to merge the data to produce a measurement versus depth log.

5.4 Surface measurements

Numerous drilling-related measurements are made at the surface (see Section 4.6, "Drilling-related measurements"). Many LWD measurements are presented versus depth, but because the downhole tool has no means to directly determine its depth, measured depth along the wellbore is acquired at the surface by determining the length of drillpipe run in the hole. Hence, all processing requiring knowledge of depth, including reindexing of downhole measurements into depth-based logs, must be performed at the surface.

5.4.1 Depth

The measured depth of a point in a well as defined by the driller is the cumulative length of drillpipe that has been run in the hole to position the bit at that point. To make this measurement each piece or joint of drillpipe is manually measured at surface with a tape measure and the specific length of each piece of pipe added to the tally as the pipe is run in the hole. No corrections for temperature, pressure, stretch, or compression are applied.

LWD and MWD depth is measured continuously using a device that determines how far the traveling block holding the drillpipe moves during drilling. At the end of drilling each joint, or several joints (stand) of drillpipe, the LWD or MWD depth measurement is adjusted to agree with the driller's depth. This generally requires applying a very small change, which is linearly distributed back along the length of the joint or stand.

Depth is measured against time. The downhole measurements are recorded against time using a clock in the tool, which is synchronized with the clock at the surface. The time index is then used to merge the measurement-time data with the time-depth data to give the measurement-depth data that is displayed on a depth log.

5.4.2 Rate of penetration

By measuring the rate of change in depth while drilling versus time, the drilling rate of penetration can be determined. The ROP is generally presented as a 5-ft- [1.5-m-] depth average (ROP5).

5.4.3 Time after bit

Each formation measurement has an associated time-after-bit measurement, which indicates how long the formation was open prior to the measurement being taken at that depth. Because this measurement is related to depth, the time-after-bit calculation for each measurement is made at the surface and then associated with the corresponding real-time and recorded-mode measurement. In the example shown in Fig. 5-2, the BHA has a GR measurement 10 ft [3 m] behind the bit and a density measurement 100 ft [30 m] behind the bit. If the well is drilled at a consistent 100 ft/h, then the GR measurement is taken 10 ft/100 ft [3 m/30 m] per hour = 0.1 h = 6 min after the bit drilled through that depth. For that same depth, the density measurement is taken when the hole has been deepened 90 ft [27 m], so that the density sensors are in front of the target depth. This takes 100 ft/100 ft/h [30 m/30 m/h] = 1 h = 60 min after the bit drilled through that point in the formation. Hence, the GR and density measurements, although acquired across the same interval, are acquired at different times and may see variations in formation properties associated with changes such as invasion or borehole degradation that occur with time.

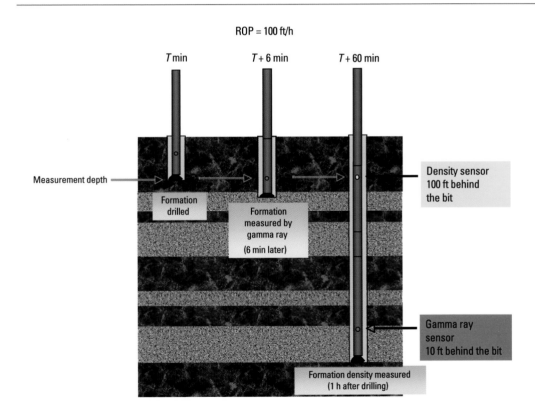

Figure 5-2. Time-after-bit example shows that the same formation depth is measured by the various sensors at different times.

5.5 Downhole data categories

LWD data can be subdivided into categories according to whether it is

- acquired relative to (indexed by) time or depth

- azimuthally focused or azimuthally symmetric

- transmitted in real time or recovered from the memory of the tool when it returns to surface.

5.5.1 Time- and depth-indexed data

All downhole measurements are initially recorded versus time. For some measurements, such as formation pressure response during a pretest at a given depth, the data remains indexed in time during interpretation. Most formation measurements, however, are time-merged with the time-depth data acquired at surface to change the measurement index from time to depth so that it can be presented as a depth log for interpretation.

A number of the drilling-related measurements are presented on both time and depth indices. Shock, for example, is relevant both as time data to show the historical performance of the BHA and the shock to which it was subjected and as depth data to show if there is a correlation between the formation type and the shock.

5.5.2 Azimuthal and nonazimuthal data

Downhole LWD measurements can be subdivided into two main groups.

Azimuthal measurements have physics that enables focusing the measurement on an azimuthal sector of the borehole, resulting in directional sensitivity. An azimuthal, or azimuthally focused, measurement has one or more directions perpendicular to the surface of a logging tool from which it receives most of its signal. Examples are density and laterolog resistivity measurements.

Nonazimuthal measurements have physics that does not enable azimuthal focusing. Rather, the measurement is acquired simultaneously from the 360° circumference of the borehole. A nonazimuthal or azimuthally symmetric measurement measures equally in all directions around the tool. Examples are propagation resistivity and neutron measurements.

5.5.3 Real-time and recorded-mode data

Data either recorded at the surface or transmitted to the surface while drilling continues is called real-time data. The transmitted data is generally a small but critical subset of the suite of measurements taken downhole. The downhole measurement data is usually depth converted, merged with the surface measurements, and presented as a log that is used to determine the formation characteristics while drilling continues. It is this data that is used for real-time decisions such as well placement.

On retrieval of the BHA to surface, the memory of the tools is recovered. This is called the recorded-mode data and it consists of all the measurements. Though there can be small differences associated with limited environmental corrections on the real-time data, the recorded-mode data should closely match the real-time data. The recorded-mode data, with a significantly greater number of data points, usually has better resolution than the real-time data.

5.6 Basic LWD log quality control

When interpreting LWD measurements there are several pieces of additional information that should be reviewed to help understand the log responses (Fig. 5-3).

The TVD curve should be presented on the log for correlating changes in the well trajectory to changes in the log responses. For example, a well that is cutting up and down across the same layer has a log response that shows several layers. The trajectory information is required to determine whether there is only one layer or many. Because the formation may not be horizontal, the correlation of updip and downdip events should not be performed based on corresponding TVD values but on log character that is seen to repeat as a mirror image. The incidence angle between the borehole and formation may be different when drilling down through a layer than when drilling up through it, so the log response may appear to be stretched (lower incidence angle) or squeezed (higher incidence angle) when comparing the mirror response with the original response.

ROP is useful for indicating how fast the BHA was moving during acquisition. Usually the ROP correlates with log features (for example, the formation is easier and hence faster to drill as the porosity increases). The ROP is typically presented as an average over the previous 5 ft [1.5 m].

The rpm indicator for the corresponding tool is required whenever an azimuthal measurement is presented. Each azimuthal tool has its own rpm indicator. As shown in the depth track of Fig. 5-3, an rpm curve that drops to zero indicates that the corresponding azimuthal sensors are not rotating across that interval and hence the azimuthal measurements are not available. Whether the tool is rotating must always be checked before attempting to interpret an azimuthal measurement. In zones where sliding occurs, the azimuthal measurement sensors may not be in good contact with the borehole wall and hence may not give valid formation measurements. An unstable, noisy, or oscillating rpm curve is a good indicator of uneven BHA rotation, usually caused by stick-slip. This behavior degrades the azimuthal measurement response, as does severe lateral shock.

Tick marks indicate the depth at which each measurement was acquired. Each measurement has its own tick marks. Widely spaced tick marks warn that the data may be undersampled over the interval. Close spacing of the tick marks, as seen in Fig. 5-3, indicates that there is sufficient data sampling to give a representative measurement of the formation.

Time after bit (TAB) indicates the formation exposure time prior to the measurement being taken at that depth, as outlined in Section 5.4.3, "Time after bit."

Shock should be presented to enable the interpreter to assess whether the BHA was moving smoothly during acquisition. Severe shock degrades measurement response and is likely to lead to downhole tool failure.

In addition to these general log quality control (LQC) indicators, there are tool-specific quality control guidelines, which are listed in each tool's Log Quality Control Reference Manual.

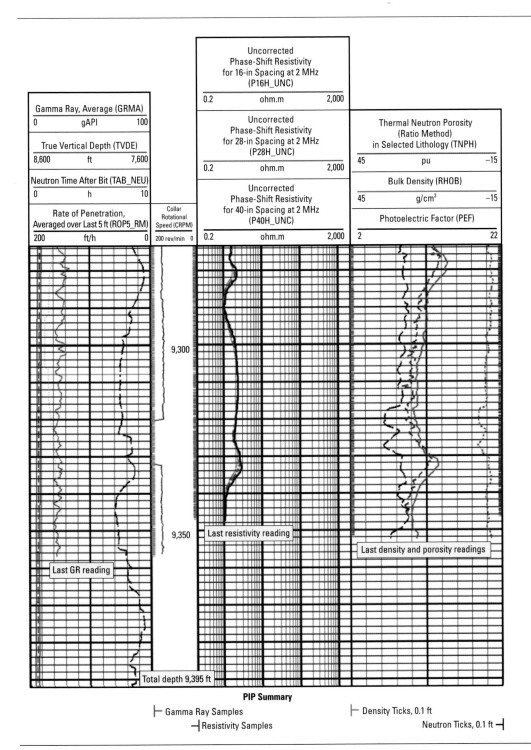

Gamma Ray, Average (GRMA)		
0	gAPI	100

True Vertical Depth (TVDE)		
8,600	ft	7,600

Neutron Time After Bit (TAB_NEU)		
0	h	10

Rate of Penetration, Averaged over Last 5 ft (ROP5_RM)		
200	ft/h	0

Collar Rotational Speed (CRPM)	
200 rev/min	0

Uncorrected Phase-Shift Resistivity for 16-in Spacing at 2 MHz (P16H_UNC)		
0.2	ohm.m	2,000

Uncorrected Phase-Shift Resistivity for 28-in Spacing at 2 MHz (P28H_UNC)		
0.2	ohm.m	2,000

Uncorrected Phase-Shift Resistivity for 40-in Spacing at 2 MHz (P40H_UNC)		
0.2	ohm.m	2,000

Thermal Neutron Porosity (Ratio Method) in Selected Lithology (TNPH)		
45	pu	−15

Bulk Density (RHOB)		
45	g/cm³	−15

Photoelectric Factor (PEF)	
2	22

9,300

9,350

Last resistivity reading

Last density and porosity readings

Last GR reading

Total depth 9,395 ft

PIP Summary

⊢ Gamma Ray Samples

⊣ Resistivity Samples

⊢ Density Ticks, 0.1 ft

Neutron Ticks, 0.1 ft ⊣

Figure 5-3. An LWD log has additional information that helps in understanding the log response.

5.7 Downhole formation measurements

Since the first generation of tools in the 1980s, the downhole capabilities of LWD tools have improved significantly to include a wider range of measurements, greater reliability, and increased real-time data capabilities.

The initial measurements were GR and simple resistivity curves, which were used more for correlation than formation evaluation. Gradually, sophisticated array resistivity, density, and neutron tools have been added to the measurement portfolio, making it possible to solve Archie's equation for formation fluid saturations using LWD data alone. Acoustic measurements such as LWD sonic and seismic services have been added, as have elemental spectroscopy, sigma (thermal neutron capture cross section), and formation pressure measurements, to further expand the use of LWD data.

The information presented to the well placement team can at times be overwhelming, so it is important to prioritize and compartmentalize the wellbore information into well placement information and formation evaluation information. Well placement decisions often require a quick response, and trying to wade through data that does not truly affect the steering decision interferes with the decision-making process. Reservoir evaluation can wait until after the steering decision is made.

Another important consideration is that LWD tools provide raw data from the wellbore, and this data must be interpreted before a decision can be made. Interpreting horizontal data can be difficult at times, so it is imperative that the proper evaluation methods be used.

5.8 Formation gamma ray measurements

Because the GR measurement is relatively unaffected by formation fluids such as water, oil, and gas, it makes an excellent correlation measurement. Where there is sufficient GR response difference between layers, the GR is sometimes used as the only correlation measurement. Because of this utility, the GR sensor is available in both MWD and LWD tools so that the measurement is available even if the MWD tool is run alone.

The GR measurement counts the number of gamma rays emitted from the disintegration of the three naturally occurring radioactive isotopes commonly found in Earth formations: thorium (Th), uranium (U), and potassium (K).

The GR can also be used to determine the proportion of clay in a formation because there is usually a higher concentration of these radioactive elements in clay. However, some uranium salts are soluble in water and can migrate dissolved in water through the formation before being deposited elsewhere. Zones enriched with uranium by this process have a higher GR reading than their clay content would normally have and hence may be mistaken as having a higher clay content than is actually the case. This situation can be resolved by using a spectral GR measurement, which distinguishes the energies of the incoming gamma rays and determines whether the gamma rays originated from a uranium, thorium, or potassium atom. The relative proportions of thorium and potassium can then be determined and the contribution from uranium removed, resulting in a more robust clay indicator.

An alternative way to estimate the formation clay content is to perform a spectroscopy measurement that determines the relative proportions of a wide range of elements (rather than just the three elements from the spectral GR), with which the clay content can be determined more directly from its elemental composition. See Section 5.11, "Lithology determination," for details.

The GR is a useful measurement for well placement for the following reasons:

- GR is relatively unaffected by fluid saturation and porosity variation. As a result, it tends to have consistent character throughout the reservoir.

- GR responds with minimal shoulder bed effect because it has a shallow depth of investigation.

- GR can resolve much finer layering in a horizontal well than can be seen on a GR log from a vertical well. There are two reasons for the improved measurement response in horizontal and high-angle wells:

  - The horizontal wellbore is at a low incidence angle to the beds. For example, a 6-in- [15-cm-] thick bed is traversed by an 8.5-in [21.6-cm] wellbore over 69.2 ft [21.1 m] of MD when the well is within 1° of formation dip (Fig. 5-4), which essentially increases its visibility to the tool.

  - The GR measurement accumulates the statistical counts from the radioactive elements in the formation over a set period of time. The slower the drilling, the more counts are acquired at a given depth. This improves the statistical precision of the measurement to develop a better representation of the formation. A vertical well can log up through the section at 3,600 ft/h [1,097 m/h] (allowing perhaps only one GR reading across a thin bed), whereas the average horizontal well is logged much slower (on the order of only tens of feet per hour). This slow penetration allows for more averaging of the GR reading and hence better precision.

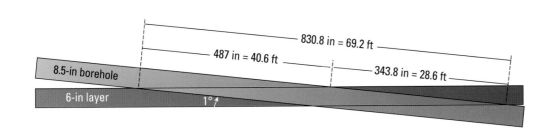

Figure 5-4. A 6-in- [15-cm-] thick bed is traversed over a measured depth of almost 831 in [2,110 cm] if an 8.5-in [21.6-cm] well intersects the layer at a 1° angle.

5.9 Formation resistivity measurements

LWD resistivity is critical for meaningful reservoir evaluation and is especially helpful when drilling in a reservoir with a fluid change such as a water contact. It can also be used for correlation, provided that the factors that could change the resistivity response of the tool are understood. For example, the deeper reading resistivity measurements can be used to detect the presence of an approaching layer of significantly different resistivity, enabling taking evasive action before the layer is encountered with the bit or detected by the shallower measurements.

LWD resistivity tools are grouped in two categories: laterolog and propagation.

Laterolog resistivity measurements are made by pushing electrical current from an electrode, across the borehole, and into the formation, with the current returning to an electrode on the tool. This laterolog current path measures the formation resistivities in series (Fig. 5-5). Hence laterolog measurements are suitable for logging in conductive muds, highly resistive formations, and conductive invasion. If the resistivity of the invaded zone, R_{xo}, is greater than R_t, the laterolog measurements will be very sensitive to the highly resistive invaded zone and not be as sensitive to the lower resistivity uninvaded formation.

Propagation resistivity measurements are derived from the changes to the phase and amplitude of an electromagnetic wave as it propagates through a formation. The propagating wave induces current loops in the formation that circle the body of the tool. These propagation-induced currents measure the formation resistivities in parallel (Fig. 5-5). Electromagnetic propagation tools work best in highly conductive formations and can operate in both conductive and nonconductive muds, provided that the contrast between the mud and formation resistivities is not so high that the induced currents tend to set up in the borehole.

5.9.1 Laterolog resistivity

Laterolog resistivity measurements are made by pushing measurement current from an emitting electrode, through the borehole to the formation, and back to a return electrode on the tool. Laterolog tools require a complete electric circuit, so they are generally restricted to operations in water-base mud.

The resistance measured by the tool is given by the voltage drop between the return and source electrodes divided by the source current:

$$r = \frac{V}{I},$$

(5-7)

where

r = resistance measured by the tool (ohm)
V = voltage drop between the source and return electrodes (volt)
I = current flowing from the source to the return electrode (ampere).

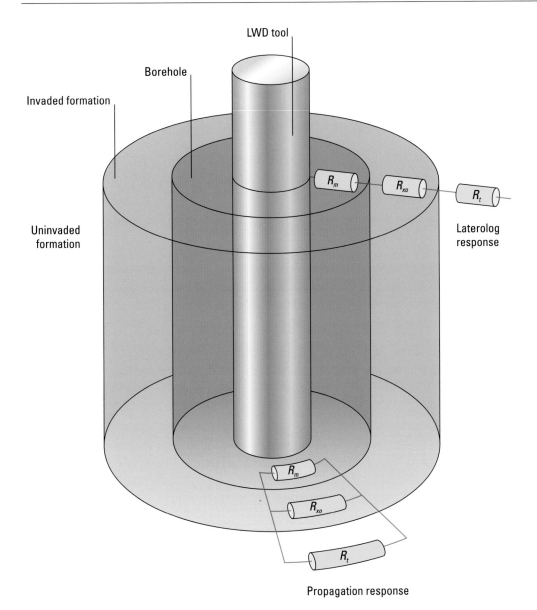

LWD tool

Borehole

Invaded formation

Uninvaded formation

R_m R_{xo} R_t

Laterolog response

R_m

R_{xo}

R_t

Propagation response

Figure 5-5. Laterolog currents measure the formation resistivities in series whereas propagation currents measure the formation resistivities in parallel.

The resistivity (expressed in ohm.m) of the formation is a property of the material, whereas the resistance (expressed in ohm) measured by the tool also depends on the volume measured. The two are related by a system constant, called the k-factor, which in simple cases is the length between the measurement electrodes divided by the area that the current passes through:

$$r = R\frac{L}{A},$$

(5-8)

where

R = formation resistivity (ohm.m)
L = characteristic measurement length (m)
A = area through which the measurement current passes (m^2).

The L/A term, which is the k-factor, is a constant for a given source-return electrode configuration.

Electrical current always follows the path of least resistance. In a homogeneous formation the current is evenly distributed around the tool, but in layered formations current "squeezes" into conductive beds, distorting the electric field. Resistive beds have the opposite effect: the current avoids them and preferentially flows in the more conductive beds. These effects, called squeeze and antisqueeze, respectively, must be taken into account in interpreting laterolog responses (Fig. 5-6).

Electrical currents can be tightly focused, making them suitable for formation imaging. Consequently LWD laterolog tools generally provide resistivity images around the borehole as well as quantitative formation resistivity data. Resistivity images are particularly useful for the identification of conductive features such as open faults and fractures because of the squeeze effect. Even fractures smaller than the imaging button size can be identified, because the measurement current takes the path of least resistance along these conductive mud-filled features. Having images from multiple depths of investigation allows evaluation of whether a feature observed on a shallow image continues deeper into the formation. Drilling-induced features such as borehole breakout can be distinguished from original formation features through the comparison of images from multiple depths of investigation.

The focusing of laterolog measurements also enables evaluating the formation resistivity at various azimuths around the borehole. Measurements from a traverse at a low incidence angle through layers can be used to assess the resistivities of the layers above and below the wellbore independently, yielding better layer definition and reserves estimates than if an average resistivity, which mixes the layer responses, is used.

The geoVISION LWD tool acquires five independent laterolog resistivities simultaneously:

- bit resistivity—the bit is used as a measurement electrode

- ring resistivity—a cylindrical electrode provides a focused laterolog resistivity

- button resistivities at three depths of investigation—azimuthally focused electrodes provide azimuthal resistivities and images.

The tool also provides azimuthal GR measurements and images.

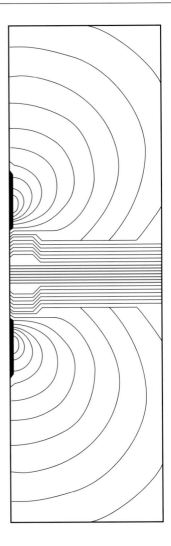 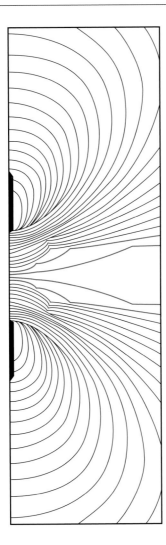

Figure 5-6. Squeeze (left) and antisqueeze (right) effects result when the measurement current (red lines) follows the path of least resistance through the more conductive layer or fracture.

Two toroidal transmitters create current flows around the geoVISION tool (Fig. 5-7). Using upper and lower transmitters creates a balanced, borehole-compensated resistivity measurement. Electrodes detect the voltages and currents resulting from the various formation resistivities. Current emitted from the bit is used to evaluate the formation resistivity at the bit. The stabilizer-sleeve-mounted shallow, medium, and deep buttons provide azimuthal, focused formation resistivity data and images. The ring resistivity is not azimuthally focused but provides a deeper depth of investigation than the button measurements.

Figure 5-7. The geoVISION measurement configuration acquires five independent laterolog resistivities simultaneously.

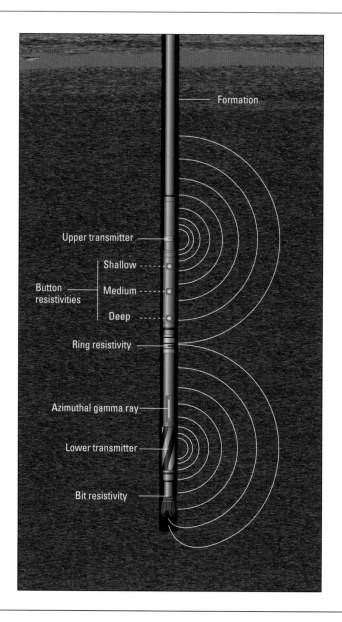

The bit resistivity measurement is useful for early detection of significant resistivity changes in the formation. Figure 5-8 shows an example of where the geoVISION resistivity-at-the-bit measurement was used to detect the resistivity change as the well penetrated the top of a high-resistivity reservoir layer from a low-resistivity shale. Drilling was stopped after only 9 in [23 cm] of reservoir penetration, and coring was able to proceed from very close to the top of the reservoir.

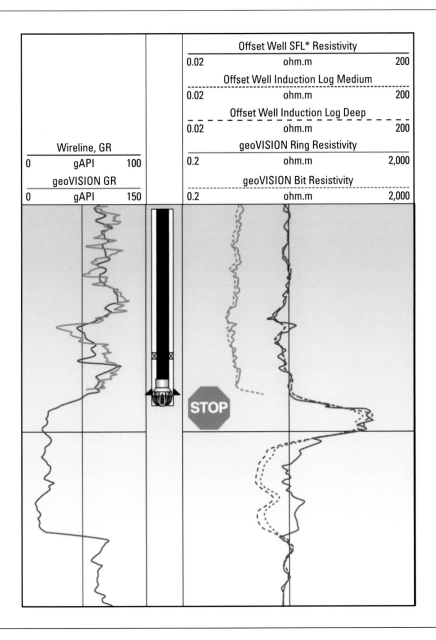

Figure 5-8. The geoVISION measurement enables early detection of significant changes in formation resistivity, as is often the case when the reservoir is reached. Casing or coring points can then be placed accurately. In this example the bit penetrated only 9 in [23 cm] into the reservoir.

Similar operations have been performed in oil-base mud because the good contact between the bit and formation allows current to flow. However, the resistivity derived in this situation is qualitative only. In nonconductive oil-base mud the current must return to the tool through contact between the drillstring and the borehole wall at an unknown location up the borehole. The resulting uncertainty in the k-factor means that the resistivity is qualitative only.

As the laterolog tool pushes current toward the bit, any conductive material below the tool, such as the housing of a motor or RSS, becomes part of the measurement electrode. Lengthening the measurement electrode degrades the vertical (axial) resolution of the measurement and moves the measure point back from the bit to the midpoint between the bit and lower laterolog tool toroid. When this occurs the bit resistivity should not be used for well placement purposes because contact of the long measurement electrode with several different resistivities simultaneously can result in complex responses. If the bit resistivity is to be used for accurate geostopping, the laterolog tool should be placed as close to the bit as possible.

The ring and button resistivities can be used only in conductive mud because they require a conductive path to the formation and back. Having azimuthally focused measurements with at least three different depths of investigation allows a simple piston invasion profile, such as that shown in Fig. 5-1, to be solved in each of the sectors around the borehole. The three unknowns of R_t, R_{xo}, and the radius of invasion, r_j, are solved using the three button measurements. In this way an invasion-corrected value of R_t can be determined in each of the sectors around the borehole for use in formation evaluation.

In the LWD laterolog in Fig. 5-9, the resistivities separate as a result of invasion in some zones and overlie in others, indicating negligible invasion. The resistivity images show features that are not visible on the GR image. Because the laterolog tool was run behind an RSS, the resolution of the bit resistivity is low as a result of the long electrode length.

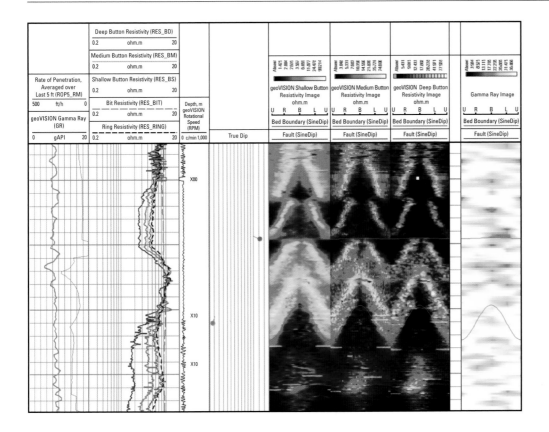

Figure 5-9. A typical LWD laterolog presentation shows the GR and ROP (Track 1); azimuthal average button, ring, and bit resistivities (Track 2); tool rpm (depth track); true dips of features picked from the images (tadpoles in Track 3); and shallow, medium, and deep button resistivities and azimuthal GR images (Tracks 4 through 7).

5.9.2 Propagation resistivity

5.9.2.1 Rock electrical properties

The impedance of a volume of rock to the passage of electrical current has a component that is independent of the frequency of the current passed through it (the conductance) and a component that depends on the frequency (the capacitance) (Fig. 5-10).

Figure 5-10. The total complex conductance of a formation is described by the two parameters of conductivity, σ, and the formation dielectric constant, ε.

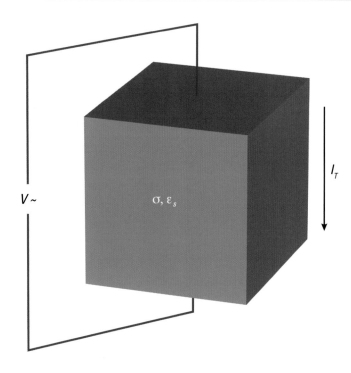

The frequency-independent component, the formation conductance, G, is given by

$$G = \sigma \frac{A}{L},$$

(5-9)

where

σ = formation conductivity (mmho)
A = area through which the measurement current passes (m^2)
L = characteristic measurement length (m).

This is the equivalent of Eq. 5-8 expressed in conductivity (the inverse of resistivity) rather than the resistivity form shown in Section 5.9.1, "Laterolog resistivity." Because laterolog tools operate at low frequencies (about 1 kHz) they are not sensitive to the frequency-dependent impedance. Propagation tools work at significantly higher frequencies (400 to 2,000 kHz), so the capacitive impedance must be taken into account.

The frequency-dependent component, the formation capacitance, X, is given by

$$X = i\omega\varepsilon_s \frac{A}{L},$$

(5-10)

where

$i = \sqrt{-1}$

ω = angular frequency ($2\pi \times$ measurement frequency)

ε_s = formation dielectric constant.

The complex conductance, Z, is given by the vector sum of the conductance and capacitance:

$$Z = G + iX,$$

(5-11)

and the total current, I_T, that flows through a formation is given by

$$I_T = (\sigma + i\omega\varepsilon)\frac{A}{L}V.$$

(5-12)

where

V = oscillating voltage applied to the formation.

The dielectric constant of a material, ε_s, is divided by the dielectric constant of a vacuum, ε_0, (8.85×10^{-12} farad/m) to create the relative dielectric constant, ε_r.

Example values of ε_r are

- ε_r of water = 80.10 (at gigahertz frequencies)
- ε_r of silicon = 11.68
- ε_r of air = 1.0054 (air behaves like a vacuum).

Because the tool response depends on both the dielectric constant and the formation conductivity, the dielectric constant must be known to estimate formation conductivity. The correlation between resistivity and the relative dielectric constant shown in Fig. 5-11 was determined for hundreds of sandstone and carbonate core samples both fully and partially saturated with water. The equation derived from fitting this data is used in the transform of the raw propagation measurements to conductivity and hence formation resistivity.

Figure 5-11. The Schlumberger correlation between resistivity and the relative dielectric constant, ε_r, was derived from hundreds of core samples.

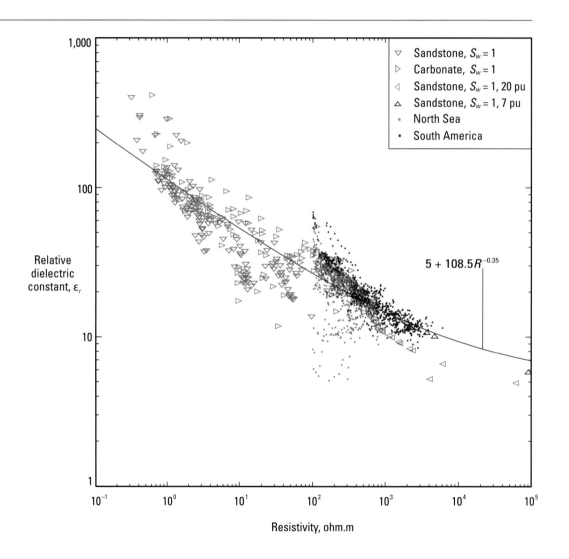

The legend of the figure:
- Sandstone, $S_w = 1$
- Carbonate, $S_w = 1$
- Sandstone, $S_w = 1$, 20 pu
- Sandstone, $S_w = 1$, 7 pu
- North Sea
- South America

$5 + 108.5R^{-0.35}$

Relative dielectric constant, ε_r

Resistivity, ohm.m

5.9.2.2 Propagation measurement physics

There are two approaches to deriving formation resistivity from the behavior of an electromagnetic wave passing through rock:

- Induction measurements use the difference in the magnetic field between two receivers that is caused by eddy currents induced in the formation.

- Propagation measurements measure amplitude and phase-shift differences between the receivers.

A wireline induction tool generates an oscillating magnetic field—typically 10 to 100 kHz—that induces eddy currents in a conductive formation. These, in turn, generate a much weaker secondary magnetic field that can be measured by a receiver coil set. Measuring the secondary magnetic field gives a direct measurement of conductivity. The higher the conductivity, the stronger the eddy currents, and the larger the secondary magnetic field. The hardware of induction tools is arranged to cancel the primary magnetic field's flux through the receiving coil set and allow measurement of the secondary magnetic field only. This is accomplished by arranging an exact number of turns and positioning of coils such that the total flux through them is zero in an insulating medium such as air. In a conductive medium the secondary magnetic field does not exactly cancel so the induction tool becomes sensitive to the eddy currents only.

If a drill collar was used to employ the same method then a similar precision for coil placement would be required. In the harsh drilling environment, a drill collar striking the wall of the borehole can easily produce 100-g_n shock, which is more than enough to ruin any precise coil positioning. LWD tools must use a scheme in which positional stability is not as demanding. This is accomplished by using a simple transmitter and receiver-pair arrangement. Precise coil placement does not matter because the phase shift and attenuation are measurable with a simple pair of coils: both quantities increase rapidly with frequency. The 2-MHz frequency has been selected as a compromise between minimizing frequency effects (dielectric) on the measured signal (low frequency preferred) while ensuring reliable phase-shift and attenuation measurements at high resistivity (high frequency preferred).

The propagation measurement is performed using loops of wire around a collar to transmit an electromagnetic wave into a formation. The difference in the phase (phase shift) and amplitude (attenuation) across a pair of coil receivers is measured (Fig. 5-12). The phase shift and attenuation are related to the formation resistivity.

Figure 5-12. The phase shift and attenuation of an electromagnetic wave propagating from a transmitter through a formation are measured at a pair of receivers.

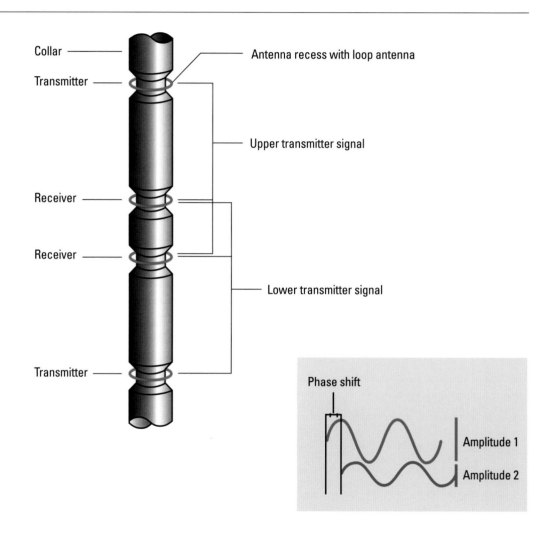

5.9.2.3 Borehole compensation

Transmitting from both above and below the receivers creates a balanced set of measurements, which compensates for effects such as hole rugosity or drifts in receiver electronics. This balancing method is called borehole compensation (BHC). Standard BHC combines data from two transmitters placed symmetrically around the receiver pair for one compensated measurement (Fig. 5-12). Schlumberger propagation tools dispense with the second transmitter, relying instead on the principle of superposition to calculate pseudo-transmitter responses for use in BHC computation (Fig. 5-13). This linear combination of three sequentially spaced transmitters provides what is called mixed borehole compensation (MBHC). The advantage is that tool length and complexity are reduced by eliminating half the transmitters that would otherwise be required. For tools using five transmitter-receiver spacings, this eliminates five transmitters.

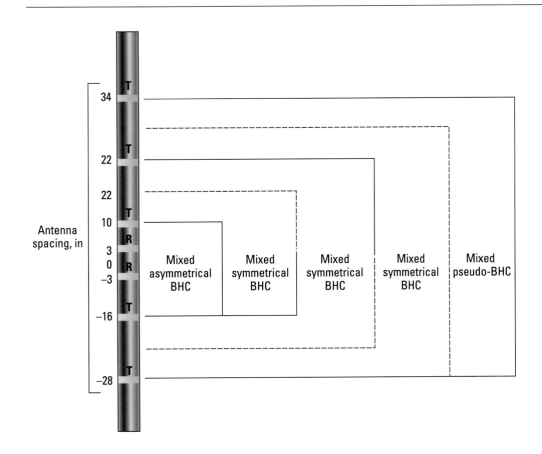

Figure 5-13. Mixed borehole compensation uses linear combinations of asymmetrically positioned transmitters to compute pseudo-transmitter responses.

Calibration of the propagation tool is performed by measuring the phase shifts and attenuations for the various transmitter-receiver spacings in a nonconductive environment (such as in air, far from any conductive material). The responses obtained in the nonconductive environment are subtracted from subsequent measurements to remove responses associated with the tool. Five MBHC phase shifts and attenuations are then transformed into five calibrated phase-shift and five calibrated attenuation resistivities using a transform similar to that shown in Fig. 5-14.

Figure 5-14. Example transforms from BHC and calibrated phase shift (top) and attenuation (bottom) to formation resistivity.

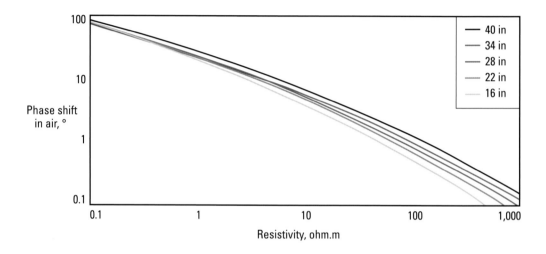

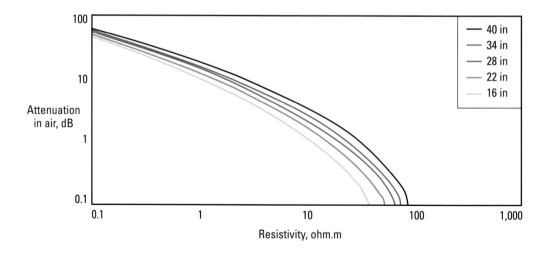

Although the transform for phase shift extends well over 1,000 ohm.m, the transform for attenuation is limited to approximately 30–70 ohm.m, depending on the transmitter-receiver spacing. This is because the difference in the wave amplitude measured between the two receivers decreases with increasing formation resistivity until the difference becomes smaller than can be reliably measured.

Because of this decrease, attenuation resistivities should not be used above approximately 50 ohm.m.

The spiky appearance of the log without MBHC in Fig. 5-15 (top) is caused by overshoots in the phase responses induced by borehole washouts and rugosity. These artifacts are canceled out by MBHC (Fig. 5-15, bottom).

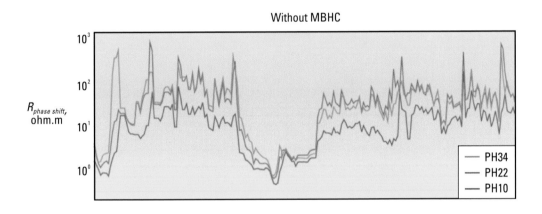

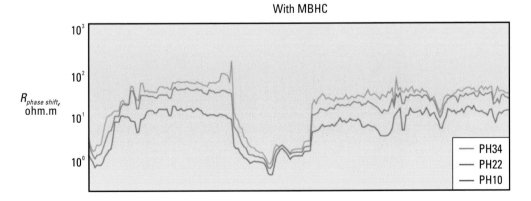

Figure 5-15. Applying MBHC (bottom) cancels out artifacts in propagation phase-shift resistivity ($R_{phase\ shift}$) logs.

5.9.2.4 Depth of investigation

The depth of investigation of a propagation resistivity measurement is controlled by four factors:

- phase or attenuation measurement of the wave
- transmitter-receiver spacing
- transmitted wave frequency
- formation resistivity.

The first three factors are controlled by tool design. The names of the Schlumberger propagation resistivity measurements encapsulate this information. For example, the resistivity measurement labeled P34H refers to the phase-shift (P) resistivity measured with 34-in transmitter-receiver spacing at high (H) frequency (2 MHz). An attenuation (A) resistivity measurement made with 22-in transmitter-receiver spacing at 400 kHz (L) would be labeled A22L.

Despite being measured on the same electromagnetic wave, the phase and attenuation measurements have independent depths of investigation. Lines of equal phase are spherical in nature because the wave travels with equal speed in all directions (Fig. 5-16, left). The corresponding phase-shift resistivity measurement is relatively shallow and axially focused. However, lines of equal amplitude form a toroidal shape around the transmitter because the amplitude is related to the energy of the wave and the tools are designed to deliver maximum energy in the radial direction. The attenuation measurement is relatively deep but less axially focused (Fig. 5-17).

Figure 5-16. Lines of equal phase around a transmitter are spherical, whereas lines of equal amplitude are toroidal.

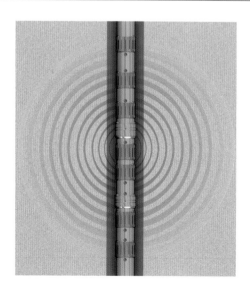
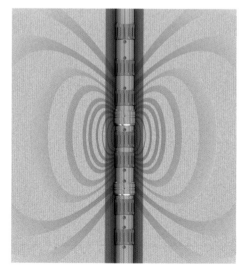

Equal-phase lines

Equal-amplitude lines

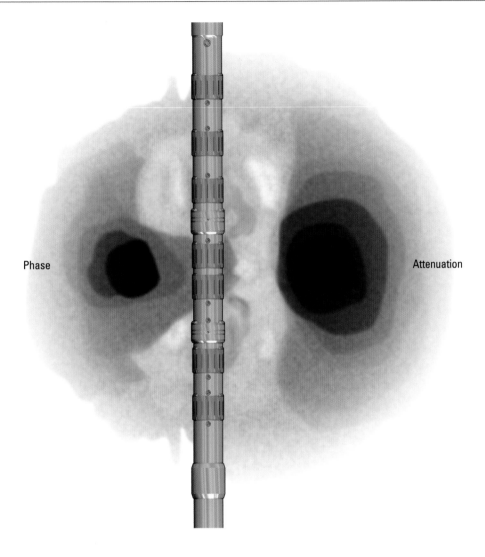

Phase

Attenuation

Figure 5-17. The phase and attenuation measurements provide two independent volumes of investigation. The phase measurement is shallow and axially focused, and the attenuation is deeper but less axially focused.

Because the depth of investigation increases as the transmitter spacing increases, the five phase-shift resistivities represent five different depths of investigation with nearly identical axial resolution. Similarly, the five attenuation resistivities represent five deeper reading measurements.

For the range of transmitter-receiver spacings in common use, all the attenuation measurements are deeper than the phase-shift measurements (Fig. 5-18).

For propagation measurements the radius of investigation is defined as the distance from the borehole at which 50% of the measurement response comes from closer to the borehole and 50% comes from deeper into the formation. Figure 5-19 shows the increased radius of investigation obtained when operating at 400 kHz.

Although the measurement depth is increased, the resistivity operating range is reduced at higher resistivities, as indicated by truncation of the attenuation response at 20 ohm.m for the 400-kHz measurements compared with 50 ohm.m for the 2-MHz response in Fig. 5-18.

Figure 5-18. The radius of investigation of both phase-shift and attenuation measurements increases with increasing formation resistivity. This plot is for a 6.75-in [17.15-cm] tool at 2 MHz.

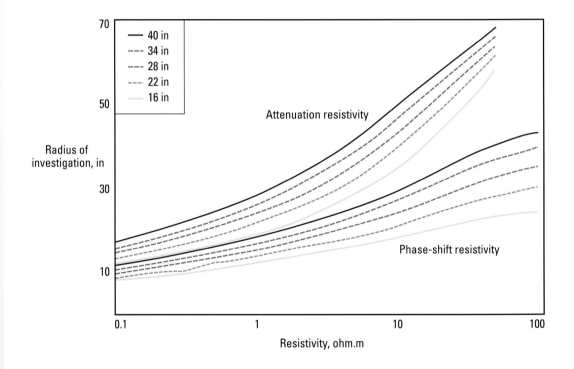

The depth of investigation of both the 2-MHz and 400-kHz measurements increases with increasing formation resistivity (Figs. 5-18 and 5-19). The measurement current induced in the formation by the transmitter seeks the path of least resistance. At low resistivities the current remains relatively close to the tool. As formation resistivity increases, the current spreads over a larger area. As outlined in Section 5.9.1, "Laterolog resistivity," the resistance of a rock is given by Eq. 5-8.

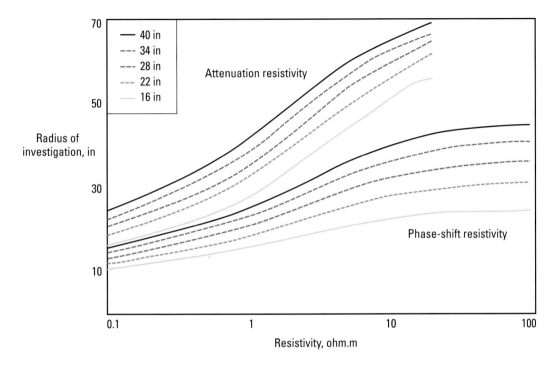

Figure 5-19. The 400-kHz radius of investigation is deeper than that of the corresponding 2-MHz measurements in Fig. 5-18. This plot is for a 6.75-in tool at 400 kHz.

For high-resistivity formations, the current spreads out over a larger area to reduce the total resistance of the path that it traverses around the tool. This spreading of the induced current results in deeper measurements with increasing formation resistivity as the current spreads deeper into the formation. It also results in reduced axial resolution with increasing formation resistivity as the current spreads along the tool.

In summary:

- Attenuation measurements are deeper than phase measurements.

- Depth of investigation increases with increasing transmitter-receiver spacing.

- Depth of investigation increases with decreasing transmitter frequency.

- Depth of investigation increases with increasing formation resistivity.

5.9.2.5 Axial resolution

The axial resolution of a measurement (also known as vertical resolution, from the days when wells were mainly vertical) is a distance that characterizes the ability of the measurement to resolve changes in the formation parallel to the tool axis.

The axial resolution of a propagation resistivity measurement is controlled by four main factors:

- phase or attenuation measurement of the wave

- receiver-receiver spacing

- transmitted wave frequency

- formation resistivity.

Figure 5-20 shows the axial response function of typical phase-shift and attenuation resistivity measurements. The axial response function can be thought of as the window along the length of the tool through which the measurement of the formation is made. The sharper the response function, the thinner the formation layer that the measurement can uniquely resolve. If a layer is thinner than the axial window, then it is averaged with the layers above and below it. The phase-shift resistivity response (top) has a thinner window (better resolution) than the attenuation measurement (bottom). This resolution corresponds with the volume of the response for the two measurements (Fig. 5-17).

There are several different definitions of axial resolution (Fig. 5-20). First, it is the interval within which a large percentage, typically 90%, of the axial response occurs (quantitative resolution). Second and most commonly quoted numerically, it is the width at the 50% point of the axial response function (width at half maximum). Third, it can refer to the smallest bed thickness for which a significant change can be detected by the measurement (qualitative resolution).

Table 5-2 lists the axial resolution, defined as the width of the axial response function at half maximum, for various phase and attenuation measurements in typical formation resistivities. As shown graphically in Fig. 5-20, phase-shift resistivities have a sharper axial resolution than that of attenuation resistivities.

The axial resolution changes only slightly with varying transmitter-receiver spacing. This is because the measurement is taken between the pair of receivers. The axial resolution is strongly dependent on the spacing of the two receivers, but because this is fixed at 6 in [15 cm] for most propagation tools, this is not a factor that needs to be considered for interpretation. The distance to the transmitter has only minimal influence on the axial resolution.

At comparable conditions, the 2-MHz measurements have sharper axial resolution than the 400-kHz measurements. In general, the 2-MHz measurements are sharper and shallower than their 400-kHz counterparts.

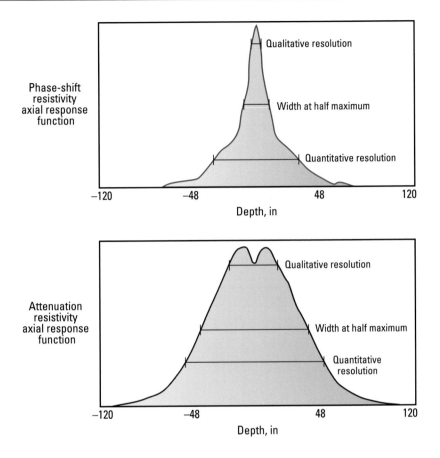

Phase-shift resistivity axial response function

- Qualitative resolution
- Width at half maximum
- Quantitative resolution

−120   −48   48   120

Depth, in

Attenuation resistivity axial response function

- Qualitative resolution
- Width at half maximum
- Quantitative resolution

−120   −48   48   120

Depth, in

Figure 5-20. The axial resolution of the phase shift-measurement (top) is sharper than that of the attenuation measurement (bottom).

Table 5-2. Axial Resolution of Various Phase-Shift and Attenuation Measurements

	Axial Resolution, ft [m]				
Transmitter-receiver spacings, in [cm]	16 [41]	22 [56]	28 [71]	34 [86]	40 [102]
R = 1 ohm.m					
Phase-shift resistivity 2 MHz	0.7 [0.2]	0.7 [0.2]	0.7 [0.2]	0.7 [0.2]	0.7 [0.2]
Phase-shift resistivity 400 kHz	1 [0.3]	1 [0.3]	1 [0.3]	1 [0.3]	1 [0.3]
Attenuation resistivity 2 MHz	1.8 [0.5]	1.8 [0.5]	1.8 [0.5]	1.8 [0.5]	1.8 [0.5]
Attenuation resistivity 400 kHz	3 [0.9]	3.5 [1.1]	4 [1.2]	4 [1.2]	4 [1.2]
R = 10 ohm.m					
Phase-shift resistivity 2 MHz	1 [0.3]	1 [0.3]	1 [0.3]	1 [0.3]	1 [0.3]
Attenuation resistivity 2 MHz	4 [1.2]	5 [1.5]	6 [1.8]	6 [1.8]	6 [1.8]

The axial resolution degrades with increasing formation resistivity. As outlined in Section 5.9.2.4, "Depth of investigation," this is because the current induced in the formation spreads out to reduce the total resistance it experiences as it circulates in the formation around the tool.

Figure 5-21 shows the 2-MHz propagation resistivity response to a 4-ft [1.2-m] layer of 100 ohm.m and another of 1 ohm.m sandwiched between 10-ohm.m layers. The phase-shift resistivities with their better axial resolution get closer to R_t in the thin beds. The shorter transmitter-receiver spacing of the 16-in [41-cm] phase-shift resistivity gets closer to R_t than the 40-in [102-cm] phase-shift resistivity because of its slightly better axial resolution. The attenuation resistivities, which have less axial resolution, read considerably lower than R_t in the 100-ohm.m layer. The axial resolution window for the attenuation resistivity includes both the 100-ohm.m layer and the 10-ohm.m layers on either side. The current induced by the tool preferentially flows in the lower resistivity 10-ohm.m layers on either side of the 100-ohm.m layer, resulting in the lower resistivity reading.

Figure 5-21. The 2-MHz phase-shift and attenuation resistivity responses differ for a 4-ft resistive (100-ohm.m) layer and a conductive (1-ohm.m) layer sandwiched between 10-ohm.m layers.

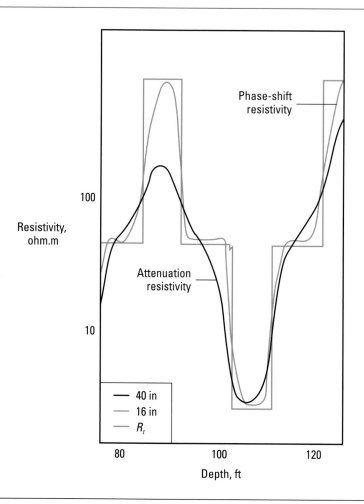

Both the phase-shift and attenuation measurements get close to R_t in the 1-ohm.m layer because it is relatively conductive compared with the surrounding 10-ohm.m layers. The measurement current, seeking the path of least resistance, preferentially flows in the low-resistivity layer.

From this example it can be seen that axial resolution effects and the path of least resistance for the induced currents within the axial resolution window must be considered in interpreting resistivity around thin beds.

In summary:

- Phase measurements have sharper axial resolution than the corresponding attenuation measurements.

- Axial resolution sharpens with increasing transmitter frequency.

- Axial resolution sharpens with decreasing formation resistivity.

- Axial resolution sharpens with closer receiver-receiver spacing (this is fixed for a tool).

- Axial resolution sharpens slightly with closer transmitter-receiver spacing.

5.9.2.6 Borehole effect

Conductive mud in the borehole immediately surrounding the tool creates an alternative path for the measurement currents induced by the transmitters. The measurement currents respond to the borehole and formation resistances in parallel. With increasing contrast between high formation resistivity and low mud resistivity, a greater proportion of the measurement current flows around the tool in the borehole rather than in the formation. Increasing borehole size also creates an easier alternative path and so increases the borehole effect.

The effect can increase or decrease the apparent resistivity response. Figure 5-22 shows the borehole corrections for the phase-shift resistivities of the EcoScope* multifunction logging-while-drilling service as a function of varying borehole diameter, mud resistivity, and formation resistivity. However, the corrections are not always in the same direction. The magnitude of the correction increases with increasing contrast between the mud and formation resistivities.

Figure 5-22. Borehole corrections for phase-shift resistivities from a 6.75-in propagation resistivity array vary with borehole size, mud resistivity, and formation resistivity.

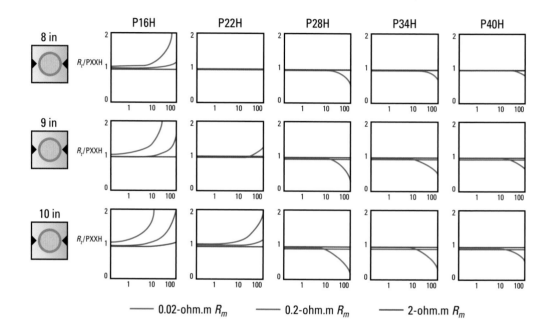

Measurements denoted with the "_UNC" qualifier (for example, P40H_UNC) have been borehole compensated (refer to Section 5.9.2.3, "Borehole compensation") but not borehole corrected. Borehole-corrected curves do not have the qualifier (for example, P40H).

In summary:

- A conductive borehole acts as an alternative path for propagation measurement currents. Borehole effect may increase or decrease the apparent resistivity response.

- Borehole-compensated but uncorrected resistivities are denoted with the _UNC qualifier.

- Borehole-corrected resistivities do not have the qualifier (for example, P40H).

5.9.2.7 Eccentricity effect

The eccentricity effect causes erratic and spiky 2-MHz resistivities when the propagation resistivity tool is run eccentered in a borehole filled with oil-base mud and surrounded by low-resistivity formation. As outlined in Section 5.9.2.4, "Depth of investigation," the 2-MHz resistivities are shallower than the 400-kHz resistivities. Low formation resistivities result in the induced currents flowing in the formation very close to the tool. If the tool is centered in the borehole, the currents continue to flow in the formation, but if the tool is eccentered, the currents may try to cross the borehole (Fig. 5-23, left). If the borehole is filled with nonconductive mud, the current is unable to traverse the borehole so its path is distorted around the borehole. This results in erratic resistivity responses as slight changes in the position of the tool in the borehole change the current path.

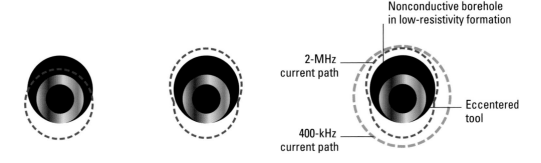

2-MHz current path

Nonconductive borehole in low-resistivity formation

400-kHz current path

Eccentered tool

Figure 5-23. An eccentricity effect occurs when the 2-MHz induced currents try to cross a nonconductive borehole (left), resulting in distortion of the current path (center), which creates erratic and spiky log responses. The 400-kHz measurement is significantly less sensitive to this effect (right).

Because of their deeper depth of investigation, the 400-kHz measurements are significantly less sensitive to the eccentricity effect. In addition, they provide better a signal-to-noise response in very low-resistivity formations. If an eccentricity effect is suspected, then only the 400-kHz resistivities should be used.

In summary, spiky 2-MHz curves may indicate an eccentricity effect in the following situations:

- The propagation tool is not centered in the borehole.

- The mud is not conductive (oil-base mud).

- The formation has a resistivity below 2 ohm.m.

In this case the 400-kHz resistivities must be used for formation resistivity evaluation.

5.9.2.8 Blended resistivities

Blended resistivities combine the best of the 400-kHz responses (at low resistivity) and the 2-MHz responses (at high resistivity) by applying simple threshold logic (Fig. 5-24).

Figure 5-24. Blended resistivities use the 400-kHz resistivities below 1 ohm.m and the 2-MHz resistivities above 2 ohm.m. Between these blending thresholds, a linear combination of the resistivities is used.

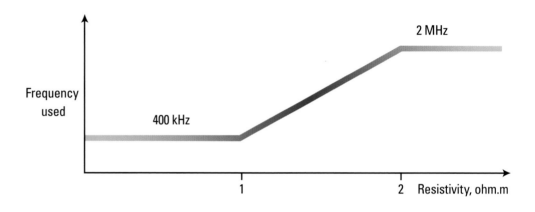

The blended resistivities curves (Fig. 5-25) are labeled in the same way as the single-frequency curves except that the final letter is B (for blended) rather than H (high for 2 MHz) or L (low for 400 kHz). For example, the P40H and P40L resistivities would be blended into a single resistivity labeled P40B.

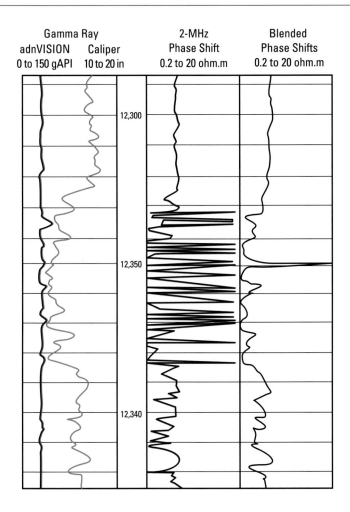

| Gamma Ray adnVISION 0 to 150 gAPI | Caliper 10 to 20 in | 2-MHz Phase Shift 0.2 to 20 ohm.m | Blended Phase Shifts 0.2 to 20 ohm.m |

Figure 5-25. Blended resistivities (right) combine the 400-kHz response at low resistivity with the 2-MHz response at high resistivity, thereby eliminating the spikes caused by the eccentricity effect seen on the 2-MHz responses (middle track) at low resistivity. The single remaining spike on the blended resistivity suggests that the blending thresholds may need adjustment.

5.9.2.9 Invasion effect

Invasion of mud filtrate into the formation near the wellbore generally results in a change in the resistivity around the well. If R_{xo} is greater than the uninvaded R_t, the condition is called resistive invasion ($R_{xo} > R_t$). If invasion reduces the formation resistivity around the wellbore, it is called conductive invasion ($R_{xo} < R_t$).

Evaluation of the invasion profile is one of the primary reasons for the development of multiple depths of investigation for resistivity tools. A greater proportion of the shallow measurement response comes from the invaded zone than for deeper measurements. Hence conductive invasion causes a spread of resistivities, with the shallow measurements reading lower resistivities than the deep measurements. This creates a conductive-invasion resistivity profile (left side of Fig. 5-26). If the invasion is very local around the borehole, the separation on the phase-shift resistivities may be significant but the separation on the deeper attenuation measurements relatively small. Deep invasion results in significant separation in both the phase and attenuation measurements.

Resistive invasion is characterized by the reverse order of the resistivities. In resistive invasion the shallowest resistivity reads highest and the deepest resistivity reads lowest (right side of Fig. 5-26).

Figure 5-26. A conductive invasion profile (left) has the shallowest resistivity reading lowest and the deepest resistivity reading highest. A resistive invasion profile (right) has the opposite order, with the shallowest resistivity reading highest and the deepest resistivity reading lowest.

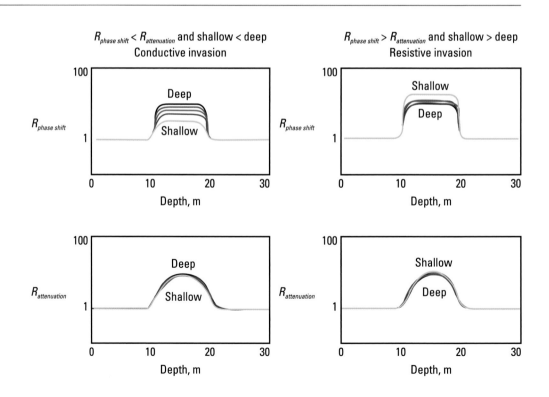

Multiple resistivities are required to solve an invasion model so that R_t can be determined. The simplest invasion model is the piston invasion profile (left side of Fig. 5-27). By using resistivities that have been corrected for borehole effects (using the known mud resistivity, R_m, and hole radius, r_h), the three unknowns R_t, R_{xo}, and r_i can be solved with the three resistivity measurement inputs.

The geoVISION tool, with three focused button resistivities, can solve a piston invasion model in each quadrant around the borehole. This enables detecting different invasion radii and determining R_t in each of the quadrants.

Propagation resistivity tools induce current loops that circulate in the formation around the tool. Hence the resistivity they provide is an average from around the borehole. When solving for a piston invasion profile with a propagation tool, it is assumed that r_i, R_{xo}, and R_t are the same in all directions around the borehole (axisymmetric).

The availability of more resistivity measurements enables solving more complex invasion profile models. The ramp profile has four variables (R_t, R_{xo}, r_{i1}, and r_{i2}) and thus requires four resistivity measurements to solve (right side of Fig. 5-27).

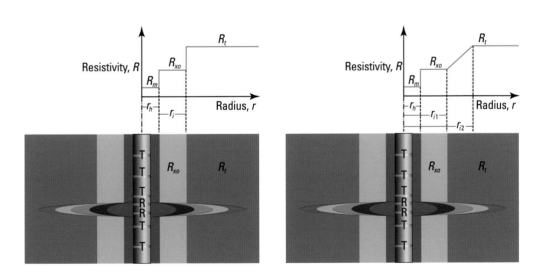

Figure 5-27. The piston or step invasion profile (left) is the simplest invasion model and can be determined with three resistivity inputs to solve for the three unknowns (R_t, R_{xo}, and r_i). Solving for the more sophisticated ramp invasion profile (right) requires four resistivity inputs to solve for the four unknowns (R_t, R_{xo}, r_{i1}, and r_{i2}). Both examples show conductive invasion. The approach is also valid for resistive invasion.

The piston invasion profile is the most common model used to interpret resistivity separations. Figure 5-28 shows modeled responses for three phase-shift and three attenuation resistivities in a piston invasion profile. A uniform 4-in [10-cm] invaded zone around the borehole with a resistivity (green line) of $R_{xo} = 1$ ohm.m is modeled in a formation where the uninvaded formation resistivity (pink line) varies from $R_t = 100$ ohm.m down to 0.3 ohm.m. Where R_t is greater than 1 ohm.m creates a conductive invasion profile. Where R_t is below 1 ohm.m creates a resistive invasion profile.

In the conductive invasion profile the shallower phase-shift and attenuation resistivities read lower than the deeper measurements because the invasion has a greater effect on them. In the resistive profile the reverse separation is observed. Although r_i remains constant the separation of the resistivities increases as the contrast between R_{xo} and R_t increases.

In summary:

- Conductive invasion—shallow phase-shift resistivities decrease. If the invasion is deep or the resistivity contrast high, the shallow attenuation resistivities can also decrease.

- Resistive invasion—shallow phase-shift resistivities increase. If the invasion is deep or the resistivity contrast high, the shallow attenuation resistivities can also increase.

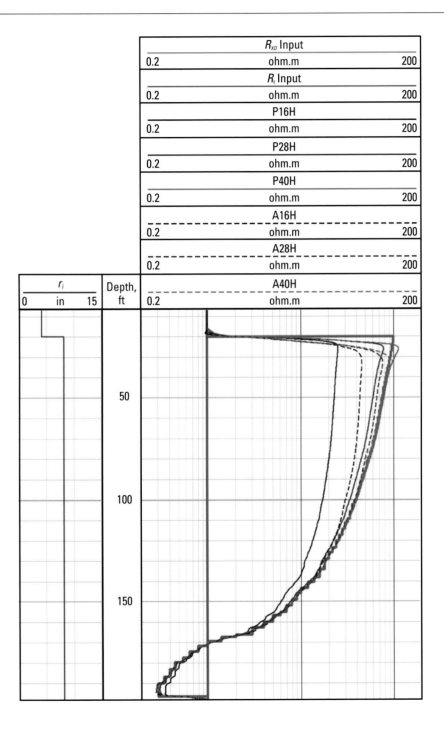

R_{xo} Input		
0.2	ohm.m	200
R_t Input		
0.2	ohm.m	200
P16H		
0.2	ohm.m	200
P28H		
0.2	ohm.m	200
P40H		
0.2	ohm.m	200
A16H		
0.2	ohm.m	200
A28H		
0.2	ohm.m	200
A40H		
0.2	ohm.m	200

r_i		Depth, ft
0	in 15	

Figure 5-28. Propagation resistivity response is modeled for conductive and resistive invasion.

5.9.2.10 Boundary effect

The boundary effect, also known as the shoulder effect, on propagation resistivity measurements is controlled by two factors:

- resistivity contrast between the layers
- incidence angle between the borehole and layer interface.

The discussion of axial resolution effects in Section 5.9.2.5, "Axial resolution," considers the effect of thin layers with differing resistivity on the propagation resistivity response when the tool is perpendicular to the layers. Boundary effects occur when the tool and layering are not perpendicular, so the induced measurement currents circulating around the tool are forced to cross the resistivity contrast rather than run parallel to it (Fig. 5-29). Proximity effects occur where the tool is not crossing the resistivity boundary but the deeper measurement currents are responding to both the local layer (in which the tool is located) and the proximate layer (the nearby layer affecting the deeper measurements).

Figure 5-29. Axial resolution, bed boundary, and proximity effects are related. Axial resolution effects are due to averaging more than one layer within the axial window of the measurement. Proximity effects are due to averaging of more than one layer within the radial depth of investigation of the measurements. Bed boundary effects are a combination of axial and radial effects as the volumes of investigation of the propagation measurements cross a change in resistivity.

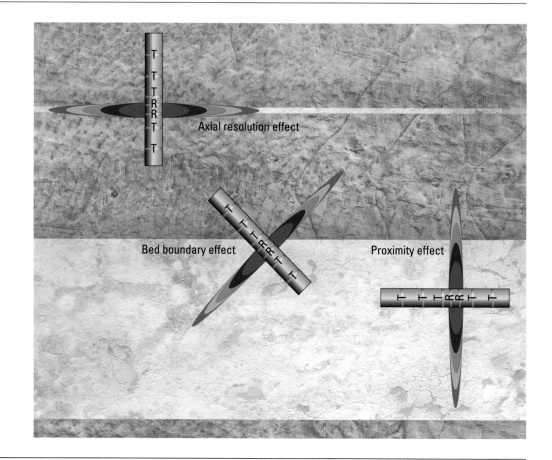

When the propagation tool and formation layering are perpendicular, the induced currents distribute themselves across the layers in inverse proportion to the resistivity. In other words, they respond to the parallel resistivity of the two layers (Fig. 5-30):

$$\frac{1}{R_{measurement}} = \frac{V_{upper}}{R_{upper}} + \frac{V_{lower}}{R_{lower}},$$

(5-13)

where

$R_{measurement}$ = apparent resistivity measured by the tool

V_{upper} and V_{lower} = volumetric influence of the upper and lower layers, respectively

R_{upper} and R_{lower} = resistivity of the upper and lower layers, respectively.

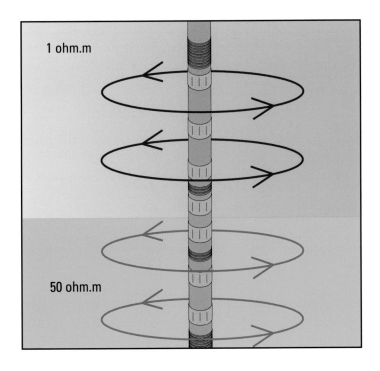

Figure 5-30. Measurement currents parallel to formation resistivity boundaries result in a resistivity response that is the volumetric parallel sum of the layer resistivities.

$$R_{measurement} = \frac{1}{\dfrac{0.5}{1} + \dfrac{0.5}{50}} = 1.96 \text{ ohm.m}$$

When a propagation resistivity tool approaches a change in formation resistivity with a low incidence angle between the tool and layering, the measurement currents are forced to cross both layers, so the resistivity response is the series sum of the resistivities (Fig. 5-31):

$$R_{measurement} = (V_{upper} \times R_{upper}) + (V_{lower} \times R_{lower}).$$

(5-14)

There is a significant difference in the resistivity response to the same layers as a function of the incidence angle between the borehole and layering.

Figure 5-31. Measurement currents forced to cross formation resistivity boundaries result in a resistivity response that is the series sum of the layer resistivities.

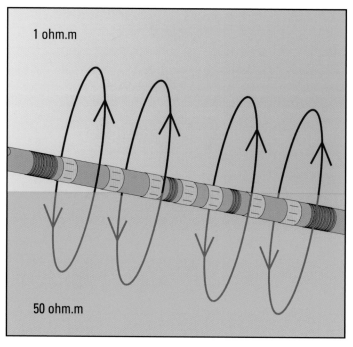

$$R_{measurement} = (0.5 \times 1) + (0.5 \times 50) = 25.5 \text{ ohm.m}$$

For propagation resistivities the measurement response is further complicated by polarization horns resulting from charge buildup at the resistivity interface. The current induced in the formation by the propagation resistivity tool must traverse both the upper and lower layers (Fig. 5-32). Kirchhoff's second law states that the voltages around a circuit must sum to zero. Given that the current is the same in the two layers but the resistance to the current is different, the voltage drop in the two layers differs. To balance this difference, charges accumulate at the boundaries. For a 2-MHz measurement, the charges are oscillating 2 million times per second with a slight delay to the measurement current, resulting in interference with the measurement signal. The interference reduces both the measured phase shift and attenuation, which results in an increase in the computed resistivities (refer to Fig. 5-14). The apparent resistivity increase as the tool approaches and crosses the resistivity contrast boundary is called a polarization horn.

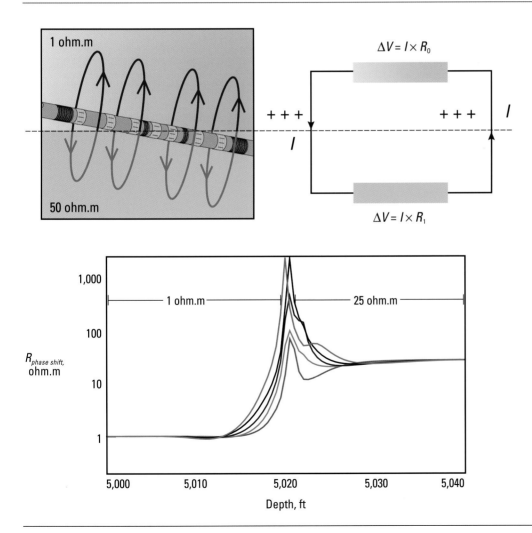

Figure 5-32. Polarization horns result from charge buildup at the resistivity interface.

The magnitude of the polarization horn depends on the resistivity contrast between the layers and the incidence angle between the tool and layers. In horizontal layers, the polarization horn magnitude increases to a maximum as the well inclination approaches 90° (parallel to the layers) (Fig. 5-33).

The polarization effect is greater on phase-shift resistivities than on attenuation resistivities. It is also greater on deeper measurements than shallower ones. This unequal effect results in a spread of the responses, from shallow reading low to deep reading high, with the phase-shift resistivities reading higher than the attenuation resistivities. The polarization effect is distinct from a conductive invasion profile, where the shallow measurements also read lower than the deeper ones but the phase-shift resistivities read lower than the attenuation resistivities.

In summary:

- Bed boundary effect results in resistivity curve separation as the measurement current crosses more than one layer.

- Polarization horns can form if

  - The incidence angle between the tool and layering forces the current to cross a resistivity contrast (approximately 45°).

  - The resistivity change is sufficient to cause significant charge buildup on the interface.

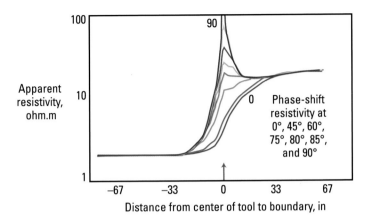

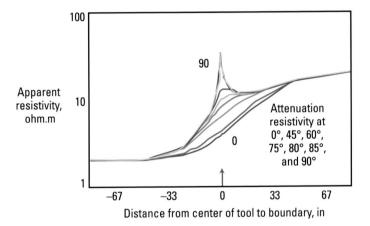

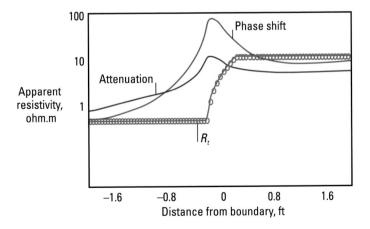

Figure 5-33. The polarization effect on phase-shift (top) and attenuation (center) resistivities increases as the tool and layering come closer to being parallel. The angles shown are wellbore inclinations to horizontal layers. The polarization effect is more pronounced on phase-shift resistivity than on attenuation resistivity (bottom).

5.9.2.11 Anisotropy effect

Resistivity anisotropy is the phenomenon of the measured formation resistivity varying depending on the direction it is measured. This is common where formations are thinly layered. The resistivity in layered formations is also called transverse isotropic (TI) because the resistivity does not vary when measuring in any direction parallel to the layering. Only when measuring across layers does the anisotropy become apparent.

In the simple bimodal layer model shown in Fig. 5-34, the horizontal resistivity, R_h, is the volumetric parallel sum of the layer resistivities:

$$\frac{1}{R_h} = \frac{V_1}{R_1} + \frac{V_2}{R_2}.$$

(5-15)

The vertical resistivity, R_v, is the volumetric series sum:

$$R_v = (V_1 \times R_1) + (V_2 \times R_2).$$

(5-16)

The terms "horizontal" and "vertical" resistivity are somewhat misleading. The quantities of interest are the resistivities measured parallel and perpendicular to the layering. The horizontal and parallel terminology is based on the assumption that the layering is horizontal, which may not be the case. To be more accurate, R_h should be called $R_{parallel}$ and R_v should be called $R_{perpendicular}$.

Figure 5-34. Resistivity anisotropy can be caused by alternating formation layers with different resistivities.

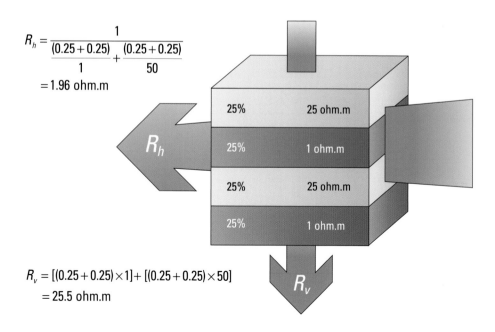

$$R_h = \frac{1}{\dfrac{(0.25 + 0.25)}{1} + \dfrac{(0.25 + 0.25)}{50}}$$
$$= 1.96 \text{ ohm.m}$$

$$R_v = [(0.25 + 0.25) \times 1] + [(0.25 + 0.25) \times 50]$$
$$= 25.5 \text{ ohm.m}$$

The propagation resistivity response to multiple thin layers can be considered the sum of numerous shoulder bed effects. Rather than a single polarization horn associated with traversing a single resistivity change, anisotropy causes the phase resistivity to read continuously higher than the attenuation resistivity. As with a polarization horn at a single boundary, there must be sufficient resistivity contrast between the layers, and the angle between the borehole and layering must be sufficiently low to force the measurement currents to traverse multiple layers (Fig. 5-35, right).

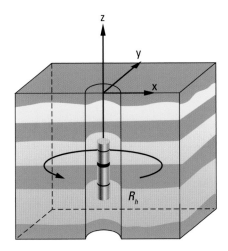

Perpendicular to layers
Propagation tools measure R_h

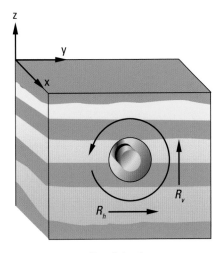

Parallel to layers
Propagation tools measure a combination
of R_h and R_v

Figure 5-35. A propagation resistivity tool oriented perpendicular to formation layering measures R_h (left), whereas a tool close to parallel with the layering measures a combination of R_h and R_v (right).

In summary:

- The tool must be at a sufficiently low incidence angle to the layering (less than 45°) to force induced currents to cut through multiple layers and thus exhibit sensitivity to the anisotropy.

- Anisotropy results in a log response similar to a polarization horn extended along the borehole (Fig. 5-36).

Figure 5-36. The resistivity log (left) shows the anisotropy signature of the phase-shift resistivities increasing from shallowest to deepest and reading higher than the attenuation resistivity. R_h and R_v were computed from the data. As with polarization horns, sensitivity to anisotropy increases as the incidence angle between the borehole and layering decreases (top right). TR = transmitter-receiver spacing.

Effect of Dip Angle on Anisotropy Response

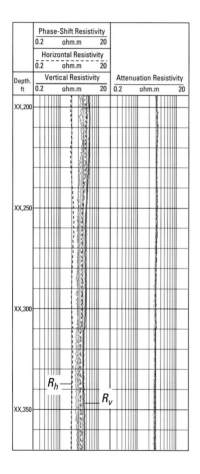

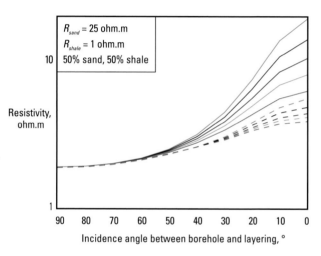

- Both phase and attenuation resistivities increase from shallowest to deepest, but the phase-shift resistivities read higher than the corresponding attenuation resistivities (Fig. 5-37).

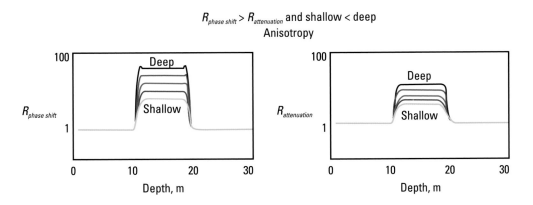

$R_{phase\ shift} > R_{attenuation}$ and shallow < deep
Anisotropy

Figure 5-37. Resistivity anisotropy causes the phase-shift resistivities to increase from shallowest to deepest and read higher than the attenuation resistivities.

5.9.2.12 Proximity effect

A proximity effect occurs when the deeper propagation measurements respond to more than one layer (Fig. 5-29). The proximity of a conductive layer is particularly apparent because the propagation measurement depth of investigation increases with the increasing local formation resistivity and the induced current seeks the path of least resistance. A conductive layer gives the induced current an alternative path with lower resistance.

In the example shown in Fig. 5-38, the well crosses from a 100-ohm.m layer into a 1-ohm.m layer. The phase-shift resistivities are unaffected by the proximity of the low-resistivity layer until the tool is very close to the layer. The deeper attenuation measurements begin decreasing from the local layer resistivity of 100 ohm.m while the lower layer is still 10-ft [3-m] TVD away. As the well approaches the 1-ohm.m layer, the attenuation resistivities decrease further as more of their volume of investigation is within the low-resistivity layer. The high resistivity contrast and low incidence angle between the borehole and layering produce a large polarization horn.

As the well enters the 1-ohm.m layer, the phase-shift resistivities rapidly drop to 1 ohm.m. The attenuation resistivities remain higher because these deeper measurements continue to have a proportion of their response from the 100-ohm.m layer above. When the well is approximately 3-ft [0.9-m] TVD from the high-resistivity layer, all the phase-shift and attenuation measurements read the local resistivity of 1 ohm.m. The depth of investigation of the propagation measurement decreases as the local resistivity decreases, and because the high-resistivity layer above does not offer a path of lower resistance to the measurement currents, they remain in the 1-ohm.m layer.

In summary:

- Responses on the deep measurements not seen on the shallower measurements can be due to proximity effects from a nearby layer.

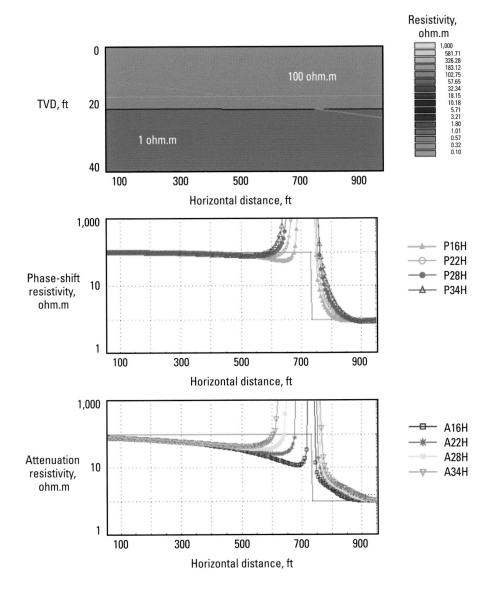

Figure 5-38. The proximity of a low-resistivity layer reduces the deeper attenuation resistivities (bottom) before the shallower phase-shift resistivities respond (middle) as the well (green line, top) traverses from high resistivity (brown) to a low-resistivity layer (blue).

5.9.2.13 Dielectric effect

As outlined in Section 5.9.2.1, "Rock electrical properties," at propagation measurement frequencies the total impedance of a rock has a direct current (DC) resistance component and a frequency-dependent component, which is a function of the formation dielectric constant. Common practice is to transform attenuation and phase shift independently to resistivity, assuming a certain transform between the relative dielectric constant and resistivity. This relationship, shown as the dielectric assumption in Fig. 5-39, loses accuracy at high resistivity. By combining the measured phase shifts and attenuations, the formation resistivity and dielectric constant can be determined simultaneously without need for a transform, as shown graphically in Fig. 5-39. The resistivity determined in this manner extends the range of measurement, typically up to 3,000 ohm.m.

Figure 5-39. Using both the measured attenuation and phase shift as inputs, the formation resistivity and dielectric constants can be determined graphically, rather than depending on the dielectric assumption. This chart is for the 2-MHz response at 40-in transmitter-receiver spacing on a 6.75-in- [17.1-cm-] diameter tool.

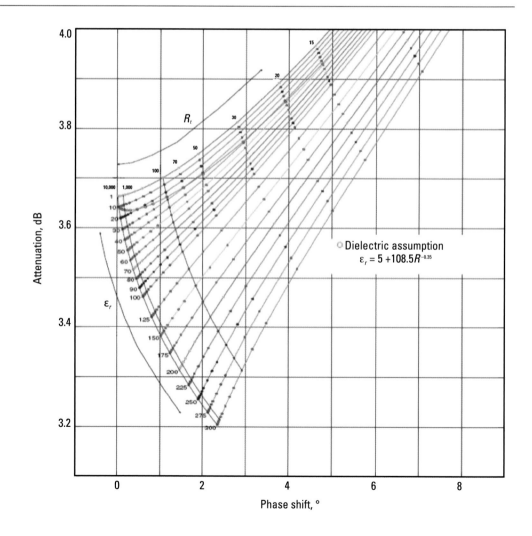

Dielectric assumption
$\varepsilon_r = 5 + 108.5 R^{-0.35}$

The attenuation measurement is more sensitive than the phase-shift resistivities to the true formation dielectric constant being higher than the formation dielectric constant calculated from the empirical dielectric
assumption equation. This tends to occur in unusual high-resistivity formations such as volcanic rocks. In these
cases the shallow attenuation measurements show a higher resistivity than the deeper attenuation measurements. The phase-shift resistivities show little or no separation (Fig. 5-40).

In summary:

- An unusually high shallow attenuation response in a high-resistivity formation may be due to the dielectric
effect.

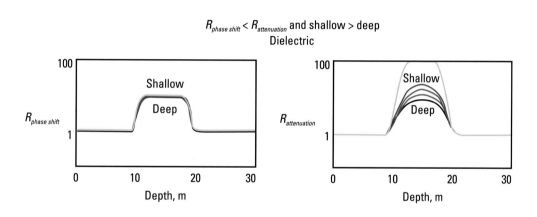

Figure 5-40. The dielectric effect
causes the shallow attenuation
resistivities to read higher than
the deep attenuation resistivities.
The phase-shift resistivities show
little or no separation.

5.9.2.14 Propagation resistivity separation summary

Real-time interpretation of propagation resistivity measurement separation is of significant benefit in well placement because it can help identify a number of formation features. Recorded-mode processing is available to assist in identifying the various causes of separation once the full data set is available. However, well placement requires real-time interpretation. Figure 5-41 summarizes the more common separations that may be observed. Each represents a single effect. Actual data may be responding to more than one effect simultaneously.

Figure 5-41. Although single effects are summarized for resistivity separation, data may respond to more than one effect. Phase-shift (blue) and attenuation (red) resistivities range from shallow (vertex) to deep (base).

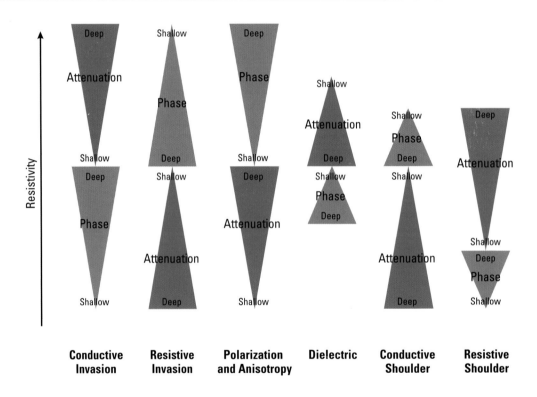

5.9.3 Resistivity tool selection

Laterolog and propagation resistivity tools both work well in many environments. In nonconductive mud, however, laterolog tools cannot give quantitative resistivities because the measurement current is unable to cross the borehole to and from the formation. In these situations the propagation resistivity is the only option. In high-resistivity formations with low mud resistivities the currents induced by the propagation tool are usually constrained to the conductive borehole, with relatively little current traversing the formation. This severe borehole effect makes the resulting resistivity unsuitable for quantitative formation evaluation. In this circumstance the laterolog measurement is the only option. These operating domains are shown schematically in Fig. 5-42.

The selection criteria assume homogenous, uninvaded formations. Both the laterolog and propagation resistivity tools are suitable for quantitative formation resistivity evaluation in the complementary region.

The selection of a resistivity tool for operations in the complementary region is based on the following:

■ sharp axial resolution required—laterolog measurement preferred

■ imaging and dip determination required—laterolog measurement preferred

■ greater depth of investigation required—propagation measurement preferred

■ conductive invasion expected ($R_{xo} < R_t$)—laterolog measurement preferred

■ resistive invasion expected ($R_{xo} > R_t$)—propagation measurement preferred.

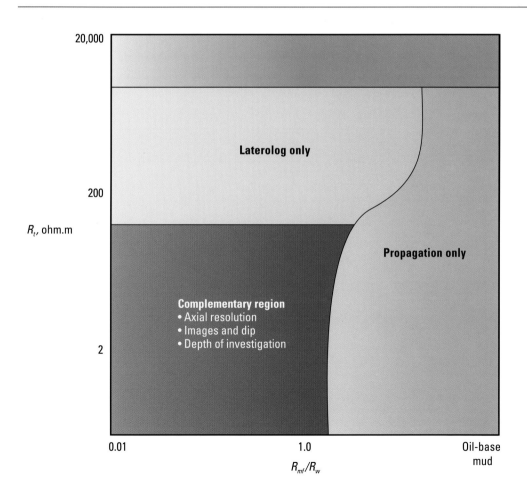

Figure 5-42. Resistivity tool selection is charted for homogeneous, uninvaded formations. R_{mf} = mud-filtrate resistivity.

As described in Section 5.9, "Formation resistivity measurements," propagation tools measure cylinders of changing resistivity around the borehole in parallel whereas laterolog tools measure them in series. Figure 5-43 repeats this important concept from Fig. 5-5.

Figure 5-43. Propagation-induced currents measure concentric formation resistivities in parallel whereas laterolog currents measure them in series.

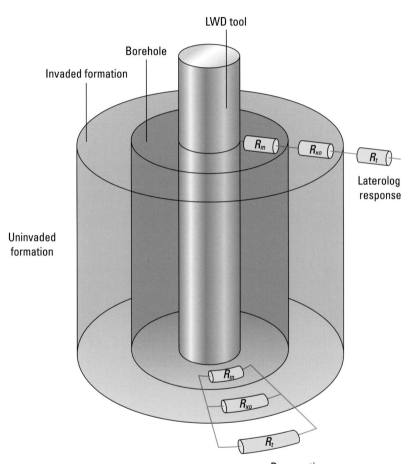

Because conductive invasion is sensed in parallel with the higher uninvaded formation resistivity by a propagation measurement, it "short circuits" the propagation measurement of R_t. The laterolog series measurement responds primarily to the high R_t, so it is preferred because relatively little invasion correction is required.

The reverse is true for resistive invasion. In this case propagation is the preferred measurement because the induced currents preferentially measure the low R_t behind the resistive invaded zone. The laterolog is severely affected by the high invaded zone resistivity in series with the uninvaded zone and reads significantly higher than R_t.

5.10 Formation porosity measurements

No single measurement directly delivers true formation porosity. Formation density is often used to derive porosity based on Eqs. 5-2 and 5-3, introduced in Section 5.3, "Petrophysics fundamentals."

The formation porosity can be determined provided that the fluid and matrix densities are known. However, they are typically unknown, so additional measurements sensitive to the fluid (neutron and magnetic resonance measurements) and lithology (neutron response, PEF, and spectroscopy) are required. Knowledge of the formation lithology enables determining the grain density.

5.10.1 Formation density

Formation density is determined based on the down-scattering of gamma ray energies as they interact with the electrons of the atoms in the formation. This is called Compton scattering, and it creates a reduction in the number of gamma rays of a specific energy range with increasing electron density of the material they pass through. Based on a correlation between the number of electrons and atomic mass, the measured electron density is converted to bulk density.

Although formation density can be determined with a single GR source and detector, in most logging tools two detectors are used to create a compensated density measurement, which corrects for parallel standoff between the detectors and borehole wall. The detector closer to the source is called the short-spacing detector; the one farther from the source is called the long-spacing detector (Fig. 5-44).

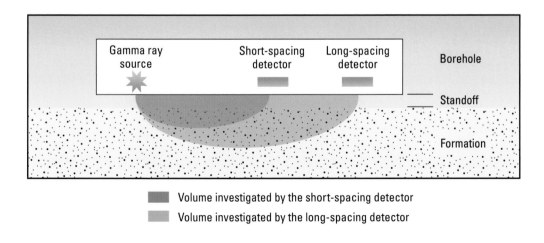

Figure 5-44. A dual-detector density measurement can compensate for the effect of limited standoff between the detectors and the borehole wall.

The spine-and-ribs correction compares the density measurements of the short- and long-spacing detectors (Fig. 5-45). If they read the same density they plot on the spine and no correction is applied. As standoff increases, the density measured by the short-spacing detector, $\rho_{short\ spacing}$, decreases more than that measured by the long-spacing, $\rho_{long\ spacing}$, because the short-spacing detector has a shallower depth of investigation and hence is more affected by the mud density, ρ_{mud}, and thickness in front of the detector. Because the short-spacing density decreases more than the long-spacing density, the data falls on a rib. The shape of the rib is characterized during tool design for determination of the density correction, $\Delta\rho$, that must be applied to the long-spacing density to recover the true formation density corrected for standoff. Because $\Delta\rho$ is an indication of standoff between the detectors and formation, large values of $\Delta\rho$ (greater than 0.2 g/cm³) indicate excessive standoff and the corresponding density data should be disregarded.

Figure 5-45. The spine-and-ribs technique for density measurements corrects for standoff between the detectors and the borehole wall.

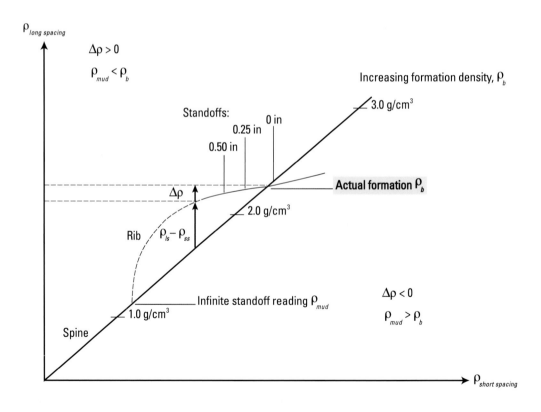

Because gamma rays are relatively easily stopped, the density measurement is shallow (2 to 3 in [5 to 8 cm]). However, it can be well focused, which makes it suitable for azimuthal measurements and imaging. Contact with the formation is enhanced by the presence of a stabilizer with special low-density windows installed to exclude drilling mud and improve gamma ray transport to and from the formation. Stabilizers can affect the drilling tendency of the BHA, so the decision to run stabilized or slick (without the density stabilizer) is a decision that should be made in consultation with the directional driller.

5.10.2 Neutron response

Neutrons are emitted from a source at high energies (millions of electron volts, eV) and lose energy as they interact with elements in the formation (Fig. 5-46).

Neutrons can be classified into three main categories according to their energy level:

- fast neutrons—energy in excess of 1,000 eV

- epithermal neutrons—energy between 10 eV and 0.4 eV

- thermal neutrons—in thermal equilibrium with their surroundings, with an average kinetic energy of about 0.025 eV at 68 degF [20 degC].

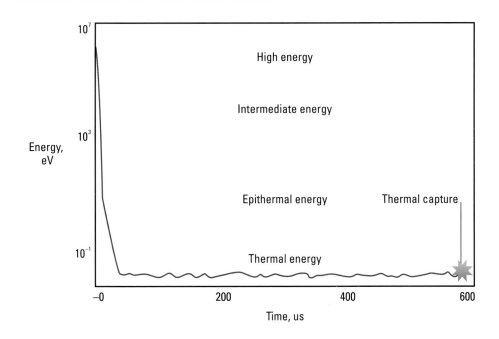

Figure 5-46. Neutrons are classified as fast (high and intermediate energies), epithermal, and thermal.

Because neutrons have no electrical charge, they pass through the electron cloud and interact with the nucleus. There are four main types of neutron interactions:

- Elastic scattering—A neutron that strikes an atom scatters and loses some of its energy to the target nucleus. The amount of energy transfer from the neutron to the nucleus depends on the mass of the nucleus (the bigger the nucleus, the less the transfer).

- Inelastic scattering—When a fast neutron strikes an atom, a portion of its energy goes into exciting the target nucleus. The target nucleus de-excites and emits gamma rays with an energy that is characteristic of the atom. The neutron is deflected and continues moving at reduced velocity.

- Radiative capture—A thermal neutron is absorbed by the target nucleus, producing a compound nucleus (isotope). The nucleus de-excites immediately, emitting gamma rays with an energy that is characteristic of the atom.

- Activation—As described for radiative capture, an atom that results in a radioactive isotope or by nuclear reaction absorbs a neutron. After a delay that is governed by the half-life of the isotope, the new atom experiences radioactive decay, emitting gamma rays and other particles.

The flux of fast neutrons continuously emitted by the neutron source travels out in all directions into the formation. As the neutrons progress, they go through three different phases:

- Slowing-down (moderation) phase—Neutrons are slowed through elastic scattering by successive collisions with atoms and lose their energy until they reach thermal equilibrium with their surroundings. Like when two billiard balls collide, maximum energy transfer occurs when the colliding particles are of similar mass. Because the nucleus of a hydrogen atom has the same mass as a neutron, it has the biggest effect in neutron moderation. Neutrons colliding with larger nuclei "bounce" back with the majority of their original energy. The probability of a collision with a given type of nucleus is proportional to the number of nuclei present per unit volume and the scattering cross section of the nucleus (which can be thought of as how big the nucleus is to a neutron). Table 5-3 lists the scattering cross sections for common formation elements. The combination of a large scattering cross section and a mass equal to that of the neutron makes hydrogen the most efficient neutron moderator.

Table 5-3. Scattering Cross Section for Common Formation Elements

Element	Atomic Number (Z)	Scattering Cross Section, barns (b)	Average Number of Collisions to Reduce Neutron Energy from 2 MeV to 0.025 eV
Hydrogen, H	1	20	18
Carbon, C	6	4.8	115
Oxygen, O	8	4.1	150
Silicon, Si	14	1.7	261
Chlorine, Cl	17	10	329
Cadmium, Cd	20	5.3	1,028

- Thermal neutron diffusion phase—Thermal neutrons continue to make elastic collisions with nuclei, but maintain a constant energy on average. On average the neutrons gain as much energy from collisions as they lose. As the name of this phase suggests, the thermal energy level depends on the temperature of the formation.

- Thermal neutron absorption phase—Eventually, each thermal neutron suffers a collision in which it is absorbed (captured). In this process a small portion of the neutron mass is converted into energy and the capturing nucleus is moved to an excited state. De-excitation generally occurs simultaneously, with the emission of one or more capture gamma rays with discrete energies that are characteristic of the nucleus involved. The probability of neutron capture is a function of energy and is low at high energies but increases as the neutrons slow down. The highest probability of neutron capture usually occurs at thermal energy. Table 5-4 lists capture cross sections for common formation elements. Boron, for example, has a high capture cross section and is used in neutron source shields. Chlorine has a relatively high capture cross section, and, given its abundance in both formation water and drilling mud, is the element primarily responsible for thermal neutron capture.

Table 5-4. Capture Cross Section for Common Formation Elements

Element	Atomic Number (Z)	Capture Cross Section, b
H	1	0.332
Boron, B	5	759
O	8	0.0003
Si	14	0.16
Cl	17	33.2
Ca	20	0.43
Gadolinium, Gd	64	49,000

The three processes of slowing down, thermal diffusion, and thermal neutron absorption occur relatively fast, so that within a few microseconds dynamic equilibrium is reached for which the total number of neutrons absorbed is equal to the number emitted by the source. Because hydrogen is the main moderator, the size of the neutron cloud is determined by the formation hydrogen index.

Hydrogen index (HI) of the fluid in the formation pores is defined as

$$HI = \frac{H \; per \; cm^3 \; in \; the \; formation}{H \; per \; cm^3 \; in \; water \; at \; 75 \; degF \; [24 \; degC]}, \tag{5-17}$$

where H is measured in grams.

Formations with a high HI result in a small neutron cloud and hence the count rate at the detectors is low. With the exception of gas, which has a low HI, formation liquids (water and oil) contain similar amounts of hydrogen per unit volume. Thus HI can be related to porosity.

The neutron response of an LWD tool can be expressed as

$$neutron \; response = \phi(HI) + secondary \; effects, \tag{5-18}$$

where
ϕ = true formation porosity.

Secondary effects refer to influences such as thermal neutron capture and differences in the neutron response to various lithologies. These must be taken into account when deriving the formation porosity from the neutron response.

The neutron response can be determined with a single neutron source and detector, but most logging tools employ two detectors to create a compensated neutron measurement, which reduces the effect of the borehole. Detectors closest to the source are called near detectors. Detectors farther away from the source are called far detectors. The ratio of the near to far count rates is used to determine the size of the neutron cloud and ultimately the formation porosity.

The transform from count rate ratio to neutron response differs from one lithology to another because the scattering and capture cross sections of various lithologies differ. Because of this variation, neutron "porosities" are quoted either as a limestone, sandstone, or dolomite neutron porosity. The appropriate lithology should be selected for interpretation.

Even after borehole compensation, borehole effect is still the major environmental correction required for the neutron response. The impact of the significant volume of hydrogen and chlorine in the borehole fluid should be corrected before the measurement is used for interpretation.

Even after the neutron response has been corrected for environmental effects, the response is primarily related to the HI of the fluid in the formation, so to derive formation porosity the corrected neutron response must be divided by the fluid HI:

$$\phi = \frac{neutron\ response\ after\ all\ environmental\ and\ secondary\ effect\ corrections}{HI\ of\ the\ pore\ fluid}. \qquad (5\text{-}19)$$

5.10.3 Density and neutron comparison

The density and neutron measurements are usually used together. The density acts as the primary porosity measurement, but the fluid and formation densities must be known to derive an accurate porosity from the density. Because the neutron responds primarily to the HI of the fluid, comparing the two measurements on a compatible scale helps diagnose whether "standard conditions" exist in the formation.

For example, a limestone-compatible scale plots the density from 1.7 to 2.7 g/cm^3 against the limestone neutron response plotted from 60 to 0 porosity units (pu), as shown in Fig. 5-47. For a 60-pu clean limestone formation of matrix density 2.7 g/cm^3 filled with fresh water with a density of 1 g/cm^3, using Eq. 5-2 gives a bulk density of 1.68 g/cm^3. To simplify the scale, this would be rounded to 1.7 g/cm^3.

Figure 5-47. Separation between the density and neutron measurements presented on a lithology-compatible scale (limestone in this example) reveals whether the formation is a clean, freshwater-filled layer of the selected lithology.

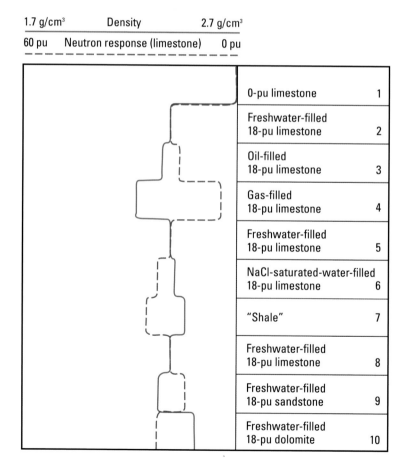

1.7 g/cm^3 Density 2.7 g/cm^3	
60 pu Neutron response (limestone) 0 pu	

0-pu limestone	1
Freshwater-filled 18-pu limestone	2
Oil-filled 18-pu limestone	3
Gas-filled 18-pu limestone	4
Freshwater-filled 18-pu limestone	5
NaCl-saturated-water-filled 18-pu limestone	6
"Shale"	7
Freshwater-filled 18-pu limestone	8
Freshwater-filled 18-pu sandstone	9
Freshwater-filled 18-pu dolomite	10

With both the density and neutron measurements plotted on a limestone-compatible scale, the two curves overlie when a clean, water-filled limestone of any porosity is encountered, as shown in the top two layer rows of Fig. 5-47.

If the fresh water is replaced by oil, the density measurement decreases because the oil has a lower density than the water it replaces. The neutron is also likely to decrease slightly because the HI of the oil can be slightly less than water. This creates the "hydrocarbon separation" in the third row of Fig. 5-47.

If the pores are filled with gas the separation becomes greater. The formation density is significantly lower because the density of gas is much lower than that of water. The neutron response is also significantly lower because the HI of gas is very low. Even though methane (CH_4) has more hydrogen per molecule than water or oil, the molecules in gas are much farther apart than in a fluid, so the hydrogen per unit volume is lower than in the liquids. These two effects result in the "gas separation" in the fourth row of Fig. 5-47.

The fifth row of Fig. 5-47 returns to the base case of freshwater-filled 18-pu limestone, where the curves overlie.

If the fresh water is replaced by salty water, a reverse separation is created. The density of water with dissolved salt is greater than that of fresh water, so the density measurement increases. The presence of salt in the water has two effects on the neutron response. The dissolved salt pushes the water molecules apart slightly, resulting in a small decrease in the HI. This would slightly reduce the neutron response. However, salt contains chloride ions, which have a very large capture cross section (Table 5-4). As previously discussed, the decreasing neutron count of the thermal neutron measurement transforms to increasing neutron response. The presence in the water of chlorides that capture neutrons means that fewer neutrons arrive at the detector to be counted. The reduced count rate drives the neutron response higher, resulting in a separation between the density and neutron responses, as in the sixth row of Fig. 5-47.

The separation that occurs in shale depends on a number of factors, which vary among the various types of shale. In general shale has a higher matrix density than limestone, resulting in the density measurement reading higher. The neutron response in shale is high owing to the presence of elements with a high capture cross section and effects on the neutrons because of the high density of the shale. Because the neutron response in shale is complicated by a number of factors, typical neutron responses are not supplied for shales in analytical charts or tables. However, a large separation caused by both the neutron response and density reading high (Fig. 5-47) is generally attributable to the presence of shale.

The eighth row of Fig. 5-47 returns to the base case of freshwater-filled 18-pu limestone, where the curves overlie.

If the formation is filled with fresh water but the matrix is sandstone rather than limestone, the separation in the ninth row of Fig. 5-47 occurs. The density decreases because the matrix density of sandstone (2.65 g/cm^3) is lower than the matrix density of limestone (2.71 g/cm^3). Because the porosity remains at 18 pu and filled with fresh water, the HI does not change from the limestone case. The change in the neutron response is due to the differing scattering and capture cross sections of the elements in the matrix. In the case of sandstone, more neutrons get through to the detectors than in a limestone of the same porosity. An increasing neutron count results in a decreased neutron response.

If the matrix is dolomite rather than limestone, the effect is reversed. The dolomite has a higher matrix density (2.85 g/cm^3) than that of limestone so the measured bulk density is increased. As with the sandstone case, the HI of the formation does not change but the dolomite matrix lets fewer neutrons through than limestone does, so the neutron response is increased in dolomite.

To create a sandstone-compatible scale, the density scale should be changed from 1.65 g/cm^3 to 2.65 g/cm^3 and the sandstone lithology neutron response displayed rather than the limestone lithology neutron response used in this example.

5.10.4 Magnetic resonance

Nuclear magnetic resonance (NMR) refers to the way that nuclei respond to a magnetic field. Many nuclei have a magnetic moment—they behave like spinning bar magnets. These spinning magnets can interact with externally applied magnetic fields, producing measurable signals. For most elements the detected signals are small. However, hydrogen, which makes up a significant component of both water and hydrocarbons, has a relatively large magnetic moment.

Before a formation is logged by an NMR tool, the protons in the formation fluids are randomly oriented. When the tool passes through the formation, the tool generates magnetic fields that activate those protons.

First, the tool's permanent magnetic field aligns, or polarizes, the spin axes of the protons in a particular direction. This process, called polarization, increases exponentially in time with the longitudinal relaxation time constant, designated as T_1.

Next, the tool's oscillating field is applied to tip the protons away from their new equilibrium position. The protons precess around the magnetic field in the same way that a child's spinning top precesses in the Earth's gravitational field. Precession occurs as a body rotating about one axis slowly rotates around a second axis. In the NMR case, this second axis is the static magnetic field (Fig. 5-48).

The initial signal received from the aligned protons after the first oscillating field pulse is proportional to the number of hydrogen nuclei associated with the fluids in the pores. This amplitude is calibrated to provide the HI of the formation.

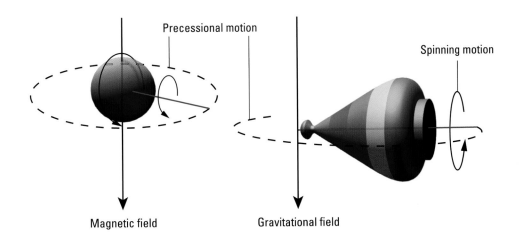

Precessional motion

Spinning motion

Magnetic field

Gravitational field

Figure 5-48. During spin precession a proton (left) spins around the blue axis, which precesses around the magnetic field (vertical). Similarly, the toy top (right) spins around the blue axis, which precesses around the gravitational field (vertical).

A sequence of pulses is transmitted by the tool to realign the protons while acquiring a series of magnetic echoes from the spinning protons. The protons begin tipping back toward the original direction in which the static magnetic field aligned them. In NMR terminology, this tipping-back motion is called transverse relaxation, designated as T_2. Relaxation results in the protons becoming desynchronized with the remaining proton population. This causes a reduction in the amplitude of the magnetic signal received by the tool's antennas. The rate of decay of the magnetic signal is the fundamental measurement of NMR logging tools and is controlled by the three relaxation mechanisms:

- surface relaxation—relaxation caused by the interaction of the protons with the walls of the pores

- bulk relaxation—relaxation caused by the interaction of protons with each other

- diffusion relaxation—apparent relaxation caused by the aligned protons moving out of the volume of investigation of the tool during the measurement sequence.

Figure 5-49. Magnetic resonance measurement involves the polarization of hydrogen nuclei followed by a sequence of realignments to measure the transverse relaxation rates (T_2) of the nuclei.

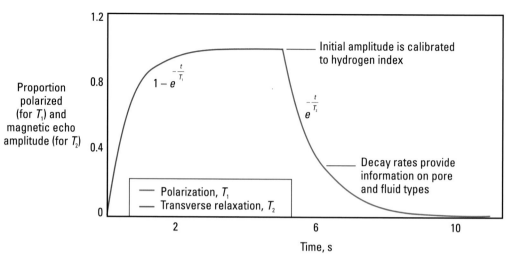

As summarized in Fig. 5-49 and Table 5-5, the magnetic resonance sequence involves an initial wait time during which the protons in the formation fluids align to the tool's permanent magnet. This polarization alignment is exponential and is controlled by the surface and bulk relaxation of the protons. Diffusion relaxation does not play a part in polarization because the oscillating field has not yet been used to create a resonant measurement volume.

The initial application of the oscillating magnetic field sets up a volume in which the protons are tipped toward the antennas, giving an initial signal amplitude proportional to the formation HI. Subsequent manipulation of the protons by using the oscillating field, which operates at the resonant frequency of the protons, enables measurement of the decay rate. Because a resonant volume is created by the oscillating field it is now possible for protons to diffuse out of the resonant volume and hence their contribution to the signal is lost. Loss of signal resulting from the movement of protons out of the measurement volume is called diffusion relaxation.

Table 5-5. Control of Relaxation of the Polarization Rate and Decay Rate

	T_1 Relaxation	T_2 Relaxation
Surface relaxation	Yes	Yes
Bulk relaxation	Yes	Yes
Diffusion relaxation	No	Yes

The T_2 decay rate measured by the tool is the cumulative sum of multiple decay rates (Fig. 5-50, left). The decay rate measured by the tool is inverted to find the component decay rates (Fig. 5-50, right). This process generates a T_2 distribution from fast decay rates on the left to slow decay rates on the right. The total area under the T_2 distribution equals the HI of the formation. The height of the distribution at a given T_2 number represents the proportion of the signal decaying at that T_2 rate.

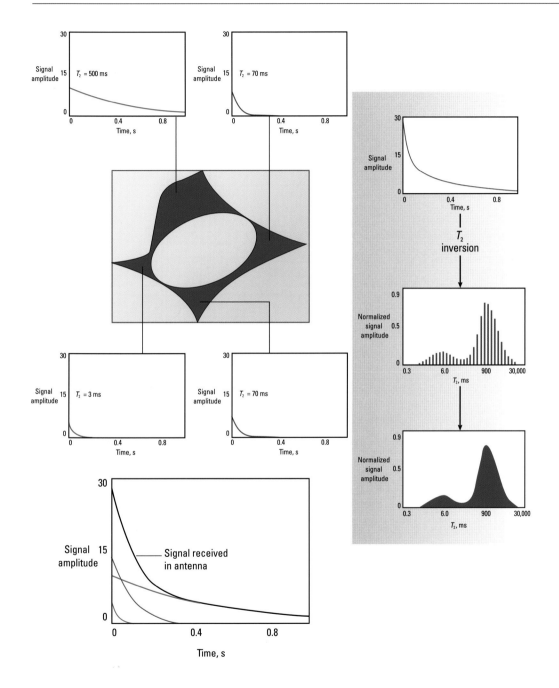

Figure 5-50. The T_2 decay measured by the tool is the cumulative sum of multiple decay rates (left). To reconstruct the array of relaxation rates, the measured decay is inverted into a T_2 distribution (right), which represents the proportion of the total signal coming from each of a number (usually 30) of different decay rates.

Because magnetic resonance logging manipulates only the protons in the fluid, the HI measurement derived from the initial amplitude is insensitive to the formation lithology (Fig. 5-51).

Figure 5-51. The magnetic resonance HI measurement (red curve on Track 2) is independent of lithology. It tracks density porosity computed for limestone in the calcite and density porosity computed for dolomite in the dolomite.

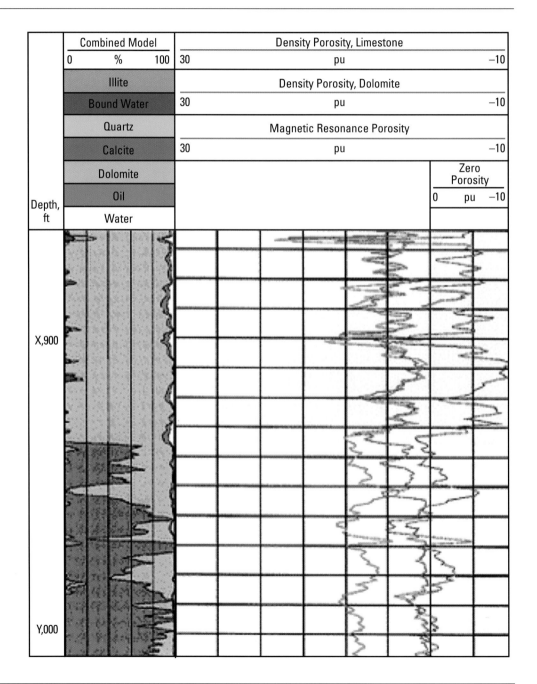

In a water-filled formation where surface relaxation is the dominant relaxation mechanism, the T_2 distribution can be related to the pore size. In small pores the protons repeatedly interact with the pore walls and hence are desynchronized, or relaxed, quickly. This causes a peak toward the low end (left) of the T_2 distribution, indicating rapid decay. In large pores the protons have fewer interactions with the walls of the pores and hence the decay is slower, with the T_2 peak farther to the right.

Magnetic resonance analysis on cores is used to determine the T_2 cutoff above which fluid flows (Fig. 5-52). Fluid below the cutoff is called bound fluid, and it is not produced when the well flows because it is either absorbed to the surface of clay (clay-bound water) or trapped by capillary pressure effects (capillary-bound fluid).

A T_2 cutoff of 33 ms is generally used in sand-shale sequences. The T_2 cutoff for carbonates is not as consistent as in sand-shale lithology because the surface relaxation coefficient (related to the paramagnetic content of the matrix) varies widely in carbonates.

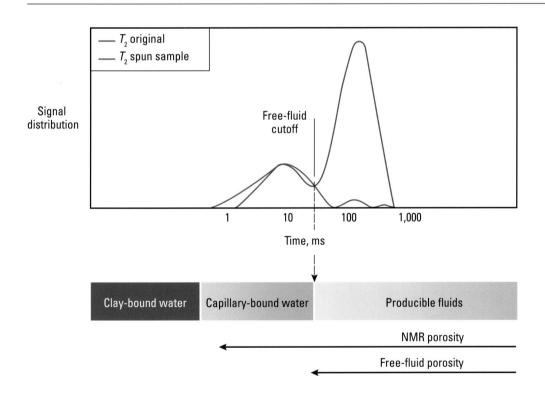

Figure 5-52. The T_2 distribution before (blue) and after (red) centrifuging a core sample shows a decrease in the T_2 components to the right that results from the free fluid being spun off and its signal lost. The nonmobile, or bound, fluid remains in the small pores. With this type of core data, a T_2 cutoff between free and bound fluids can be determined.

Although NMR tools do not directly measure formation permeability, the pore-size information is useful in transforms that correlate the T_2 distribution to permeability measured on cores. To quantify the position of the T_2 distribution, either the ratio of the bound fluid to free fluid or the logarithmic mean of the distribution (T_2 log mean, T2LM) is used. For well placement purposes, the key element is that the farther the T_2 distribution is to the right, the larger the pores and the higher the permeability.

An example[‡] of the use of magnetic resonance information for geostopping is shown in Fig. 5-53. In this Middle East carbonate the total porosity does not change significantly across the reservoir, as indicated by the relatively constant density (red), neutron (blue), and total magnetic resonance porosity (black) responses in Track 2. However, the size of the pores reduces on the flanks of the reservoir, which significantly reduces the permeability to the point that the formation is no longer commercially productive. Drilling is stopped so that valuable rig time is not wasted extending the well into nonproductive formation. The transition from one pore-size distribution to another can be seen in the shift of the T_2 distribution in Track 3 to the left with increasing depth. Based on a T_2 cutoff determined from core measurements, the total porosity is partitioned on Track 2 into free fluid (green shading) and bound fluid (yellow shading). The corresponding decrease in formation permeability with depth can be seen on the two permeability transforms in Track 1.

[‡] Rose, D., Hansen, P.M., Damgaard, A.P., and Raven, M.J.: "A Novel Approach to Real Time Detection of Facies Changes in Horizontal Carbonate Wells Using LWD NMR," *Transactions of the SPWLA 44th Annual Logging Symposium*, Galveston, Texas, USA (June 22–25, 2003), paper CCC.

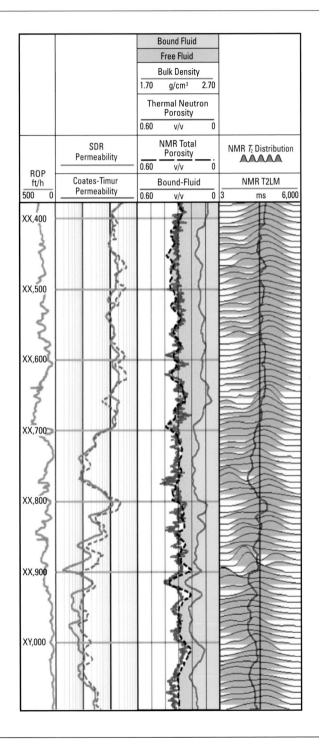

Figure 5-53. NMR logging was used to guide geostopping based on the reduction in pore size and resulting significant reduction in permeability.
Courtesy of D. Rose.

Figure 5-54 shows a sequence of formations with their corresponding magnetic resonance responses. The shale at the bottom of the formation column on the left has small pores and hence strong surface relaxation. This condition results in fast polarization and echo decay of the water in the shale. Consequently, the T_2 distribution is to the left of the T_2 cutoff, indicating that the fluid is not free to move.

The water sand has both small and large pores that are water filled. The water in the large pores may not interact with the pore walls and decay is at the bulk relaxation rate for water. Water bulk relaxation is caused by hydrogen protons in the water interacting with other hydrogen protons in other water molecules. The water in the small pores has significant surface relaxation and polarizes and decays quickly. The spread of the T_2 distribution for the water sand indicates the range of relaxation rates. Some of the distribution lies above the T_2 cutoff, indicating that some of the fluid (in the center of the large pores) is free to move. Water in the small pores and around the surface of the large pores is held in place by electrochemical and capillary effects and does not flow, as indicated by the proportion of the T_2 distribution below the T_2 cutoff.

In the oil sand the free water has been displaced by oil. The presence of the oil in the water-wet pores results in the water being pushed out to the edges of the pores. Hence the water interacts more frequently with the pore walls and is more affected by surface relaxation. The faster polarization and decay shown by the blue curve in the middle panel reflect the additional surface relaxation that the water experiences. On the T_2 distribution the water is to the left of the cutoff line. Because the oil floats in the center of the pores, surface relaxation does not have a large effect. The oil decays at its bulk rate, which is controlled by the length of the hydrocarbon chains and hence is related to the hydrocarbon viscosity. Oil consists of a wide range of molecules of different sizes, so the frequency of interactions between protons varies widely. The protons in the hydrogen of short-chain (light) hydrocarbons do not interact as frequently as those in long-chain (viscous) hydrocarbons, where the chains wrap around each other. This variation in bulk relaxation creates a range of decay rates related to the oil viscosity. Less viscous oils decay slowly and hence plot on the right of the T_2 distribution, whereas the longer chained hydrocarbons decay more rapidly because of the additional bulk relaxation and plot on the left on the T_2 distribution. In extreme cases such as tar or bitumen, the T_2 decay rate can be so fast that current tools are unable to measure the decay rate. This results in NMR porosity that is lower than the density or neutron porosities. This porosity deficit is often used as an indicator of very viscous hydrocarbons.

The gas sand is similar to the oil sand, with the oil replaced by gas. The water is displaced to the edges of the pores and undergoes significant surface relaxation. The gas in the center of the pores does not interact with the pore walls. Because gas molecules are generally small and widely spaced, they do not interact frequently with each other, resulting in low bulk relaxation effects. With low surface and bulk relaxation, the polarization of gas is considerably slower than that of liquids. The polarization "wait time" used in magnetic resonance acquisition may not be sufficient to fully polarize the gas, as shown at the top of the center panel, where the polarization of the gas (red curve) does not reach the same amplitude as the water (blue curve) despite equal volumes in the pores. The underpolarization of the gas results in a magnetic resonance porosity deficit. The NMR porosity reads lower than the density and neutron porosities in the presence of gas that has not been completely polarized. The decay rate of the gas in the center of the large pores is strongly affected

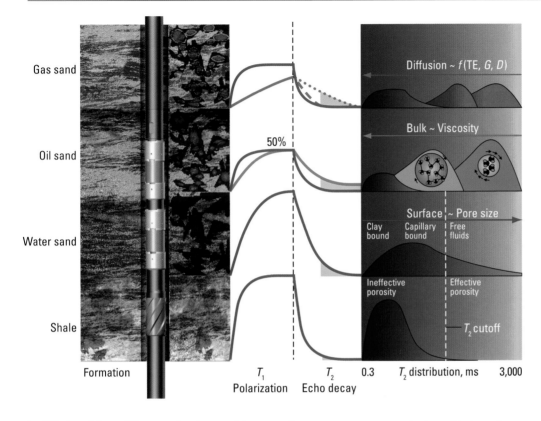

by diffusion of the mobile gas molecules out of the magnetic resonance measurement volume. The loss of the gas molecules results in more rapid decay of the signal than if the gas molecules had remained in the measurement volume.

The rate of diffusion is controlled by three factors:

- Spacing between echoes (TE)—The farther apart the echoes are spaced, the more time is available to the molecules to diffuse out of the measurement volume between echoes.

- Magnetic gradient from the measurement tool (G)—The volume of the resonant measurement shell around the tool depends on the gradient of the magnetic field in which it operates. A low gradient field creates a relatively thick shell with a large volume to surface ratio. A high magnetic gradient results in a thin shell with more surface per unit volume. Protons are more likely to diffuse out of the thin shell than the thick one. The thin shell has enhanced diffusion sensitivity.

- Diffusivity of gas (D)—The more mobile the molecules are, the greater the diffusivity of the material. Gas has a higher diffusivity than water because the gas molecules are more mobile. Water is a small molecule so it has a higher diffusivity than oil.

Magnetic resonance information enhances density-neutron porosity determination through the addition of lithology independence, enhanced fluid typing, pore-size estimation, and the associated estimation of permeability. For well placement, magnetic resonance data can be used in real time to identify tar zones and reductions in permeability for avoidance. Zones of higher permeability can be identified, and the estimation of permeability along the well enables estimating the productivity of the well during drilling.

5.10.5 Sonic logs

LWD sonic tools transmit an acoustic pulse that propagates through the formation. The energy is reflected, refracted, and converted through interactions in the borehole, formation, and fluid/solid interface, resulting in a wave train comprising various arrivals (Fig. 5-55).

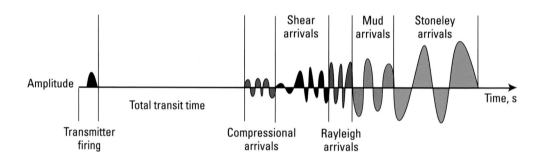

Figure 5-55. The acoustic arrivals of the sonic pulse that are of greatest interest are the formation compressional, shear, and Stoneley arrivals.

An array of equally spaced receivers measures the acoustic wave train (Fig. 5-56). The distance between the receivers is known, so measuring the time difference between an arrival at one detector and the next allows the acoustic traveltime per unit distance (in units of seconds per meter), otherwise known as the slowness, to be determined. The inverse of the slowness is the acoustic velocity through the formation (in meters per second).

The wave train comprises various arrivals for which the velocity of each can be determined by measuring arrival time at each of the detectors. The calculation process is called slowness-time coherence (STC). The coherence of the waveforms is calculated (effectively, points of greatest similarity between the waveforms are identified) and plotted as a color (red indicates highest coherence, blue lowest) on a slowness versus time crossplot (lower left in Fig. 5-57). Time refers to how long the arrival takes to propagate from the transmitter to receiver. Slowness refers to the inverse of the velocity across the receiver array. The slower the arrival propagates through the formation, the longer it takes to get to the receivers. This logic allows processing limits to be applied (white dashed lines) to exclude computing physically unlikely velocities from any outlying coherence peaks.

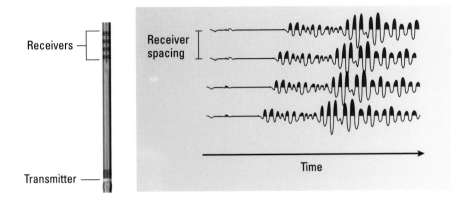

Figure 5-56. Multiple receivers measure the wave train as it passes.

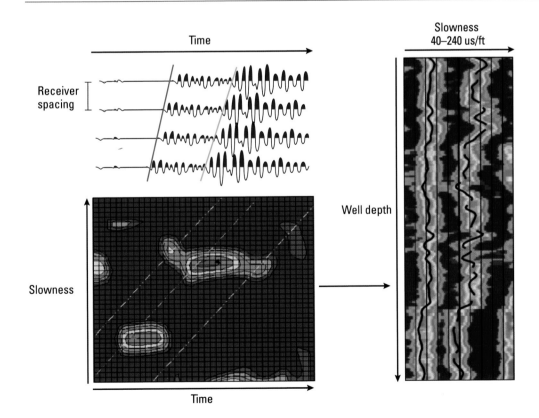

Figure 5-57. STC processing extracts the velocity of the individual arrivals at each depth. The data is then presented as slowness versus depth with log colors shading from blue (low coherence) to red (high coherence).

STC processing enables computing the compressional, shear, and Stoneley velocities. Figure 5-57 shows how the coherence (red to blue shadings) and the detected arrival peaks (black lines) are plotted on a depth log.

The compressional velocity is used for time-to-depth correction of the seismic data and porosity determination. The higher the porosity, the slower the compressional arrival through the formation. Gas has a lower acoustic velocity than oil, which is in turn slower than water. The acoustic velocity of the fluid in the pores must be taken into account when computing the porosity.

An interesting feature of porosity derived from acoustic logs is its insensitivity to isolated porosity. If a pore is unconnected to its neighbors, it is surrounded by solid rock. The acoustic wave propagates through solid material faster than through liquid, so the velocity measured at the receivers is unaffected by unconnected pores. If the pore is connected, then the acoustic energy is slowed because it has to cross the fluid in the throats connecting the pores. Acoustic porosity shows a deficit to a nuclear porosity such as density-derived porosity (which measures all porosity, whether it is connected or not). The difference between them is related to the isolated porosity. Isolated porosity affects nuclear and resistivity logs but does not allow flow, creating interpretation complexity if it is not quantified.

The compressional and shear velocities are used for calculating rock mechanical properties, which have wide application in optimizing drilling, ensuring hole stability, and inputting to mechanical earth models (MEMs) for reservoir behavior under changing pressure and stress regimes.

Stoneley velocities can be used for permeability estimation.

5.11 Lithology determination

5.11.1 Photoelectric effect

When low-energy gamma rays interact with the electrons in an atom, the gamma rays can be absorbed and an electron ejected in a process known as photoelectric capture. The probability of a gamma ray undergoing photoelectric capture is dependent on the photoelectric capture cross section of the atom and the energy of the gamma ray. The lower the energy of the gamma ray, the more likely it is to undergo photoelectric capture. The photoelectric capture cross section is dependent on the average atomic mass of the material encountered.

The average atomic mass is related to the lithology of the formation. The photoelectric factor is a measure of the GR capture cross section (units of barns per electron) of a formation. To convert to a volumetric quantity, the PEF can be multiplied by the electron density of the formation to create the volumetric photoelectric factor, U. Values of both the photoelectric factor and its volumetric equivalent for common lithologies are listed in Table 5-6.

Table 5-6. Photoelectric Factor and Volumetric Photoelectric Factor for Common Lithologies

Lithology	Photoelectric Factor, b/electron	Volumetric Photoelectric Factor, b/cm³
Quartz	1.8	4.8
Calcite	5.1	13.8
Dolomite	3.1	9
Anhydrite	5.1	15
Illite (clay)	3.5	8.7
Lignite (coal)	0.2	0.24

Gamma rays are relatively easily stopped by dense material. This is particularly true of the low-energy gamma rays involved in the photoelectric effect. The PEF measurement is the shallowest of all LWD measurements and is highly sensitive to borehole and mud conditions. The measurement is not borehole compensated like the density measurement but is made with a single detector (the short-spacing GR detector in the density section). In light mud weight with good borehole conditions, the PEF measurement can be used for lithology identification. However, in rugose boreholes or where heavy mud is used, careful quality control is required to ensure that the PEF reading is not affected by the borehole environment.

The PEF is an important indictor of formation lithology. Knowing the lithology is required so that the appropriate matrix density can be used in calculating the formation porosity from the density measurement.

5.11.2 Neutron capture spectroscopy

Once a high-energy neutron has lost the majority of its original energy through interactions with the atoms in a formation, it reaches thermal energy level. Eventually, many of the neutrons are captured by various elements in the formation, which then become a different isotope of the same elements through the addition of the neutrons to their nuclei. To convert back to a more stable energy state, some of the energy from this process is immediately released through the emission of a set of gamma rays, which have energies characteristic of the element from which they were released. By measuring the gamma ray energy spectrum emitted by a formation after bombardment with high-energy neutrons, the elemental composition of the formation can be determined.

The capture spectroscopy measurement consists of a sequence of events:

1. emission of fast neutrons from a neutron source

2. slowing down of fast neutrons to thermal energies as a result of collisions with nuclei in the formation and borehole

3. capture of thermal neutrons by atoms in the formation and borehole, and the subsequent emission of one or more high-energy gamma rays

4. detection of the capture gamma rays

5. spectral fitting of the measured pulse height spectrum to obtain the relative contributions of the gamma rays from the various elements to the total spectrum (elemental yields).

The production of gamma rays from neutron interactions involves one of two types of reactions:

- high-energy inelastic reactions

- thermal capture reactions.

For high-energy inelastic reactions, a high-energy (several MeV) neutron interacts with a nucleus in the borehole or the formation. As a result, the nucleus emits one or more gamma rays that are characteristic of the particular element (isotope). Inelastic gamma rays are important mainly for carbon/oxygen (C/O) logging. For thermal capture reactions, a slow (thermal) neutron is absorbed by a nucleus in the formation, borehole, or tool. Although neutron absorption is unlikely at high neutron energies, it occurs quite readily at thermal energies, especially in the presence of thermal neutron absorbers such as chlorine, boron, and gadolinium. When neutron capture occurs, the nucleus is often left in an excited state. When the nucleus de-excites from an excited state to its ground state, high-energy gamma rays are emitted. These gamma rays, referred to as prompt capture gamma rays because they are emitted immediately after neutron capture, are characteristic of the isotope that captured the neutron. Many of the gamma rays emitted are very energetic and scatter throughout the formation and wellbore, resulting in a cascade of associated lower energy gamma rays.

When struck by an incoming gamma ray, the detector material "scintillates," which means that it generates light, which is extracted from one end of the crystal. The scintillation light is absorbed by the photocathode material in the photomultiplier tube. The photoelectric emission of an electron is followed by electron multiplication to increase the magnitude of the signal. The result is a composite gamma ray energy spectrum, which provides a measurement of the contribution of various elements in the tool, borehole, and formation (Fig. 5-58).

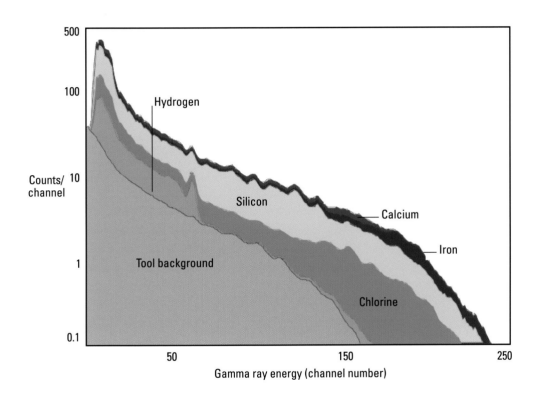

Figure 5-58. Neutron capture spectroscopy is used to determine the elemental composition of a formation by examining the prompt gamma rays emitted after neutron bombardment.

The measured spectrum is the linear summation of gamma ray signals from elements in the formation, borehole, and tool. The fraction of the spectral area resulting from gamma rays from a particular element is called the relative elemental yield. In most well logging situations, the main contributions to the measured capture gamma ray spectra come from the following elements: H, Cl, Si, Ca, Fe, S, Ti, Gd, K, Mg, Cr, Ni, and Ba. Radiative capture by these elements results in the emission of characteristic gamma rays unique for each element. The presence of an element is inferred when the characteristic gamma rays are observed in the measured spectrum.

For a specific tool and detector configuration, each element has a characteristic distribution of gamma rays called an elemental standard spectrum. To determine the presence and concentration of each element, the measured spectrum is decomposed into the relative proportions of the various elemental standard spectra by a spectral fitting procedure that uses weighted least-squares regression. This spectral stripping process delivers the relative yields of the various elements in the tool, borehole, and formation. The elemental composition of the tool is known, so the tool background contribution can be subtracted (gray region in Fig. 5-58), leaving the relative yields of the elements in the formation.

The quantities derived from spectral processing are called relative elemental yields. They can be used to determine the ratio of one element to another in the formation (for example, Ca/Si) but do not directly deliver the absolute concentration or weight fraction of any element. The next step is to convert the relative elemental yields to elemental concentrations that can then be used for quantitative interpretation. This is accomplished by using an oxides closure model, which takes its fundamental justification from the knowledge that in sedimentary rocks, most elements exist in their oxide forms and the sum of the rock-forming oxides is 1. In terms of the elements, this is expressed as

$$SiO + TiO_2 + Al_2O_3 + Fe_2O_3 + MgO + CaO + Na_2O + K_2O + CO_2 + P_2O_5 + H_2O+ + SO_2 = 1. \qquad (5\text{-}20)$$

Although only a subset of these rock-forming elements is available in the form of relative yields from the capture spectroscopy measurements, it is possible to adapt this equation to an oxides closure model that ultimately enables conversion from yields to concentrations. Several factors must be taken into consideration. First, unmeasured elements must be accounted for by relying on natural correlations between elements in sedimentary rocks. For example, for environments where aluminum can not be measured directly, an empirical relationship was developed to compute Al from the elements silicon, calcium, and iron. The complementary nature between these elements was discovered by examining a large database of chemistry and mineralogy of sedimentary rocks.

$$Al_{derived} = 0.39(100 - 2.139Si_{measured} - 2.497Ca_{measured} - 1.99Fe_{measured}). \qquad (5\text{-}21)$$

Second, because the input measurements are relative yields, the measurement sensitivity of each element must be accounted for. Some elements are more easily detected than others. Each tool has sensitivity coefficients that quantify how easily each element is detected. For one spectroscopy tool, calcium has a sensitivity coefficient of about 1.6 compared with 1 for silicon. This means that a calcium atom can be detected 60% more easily in the spectra than a silicon atom.

Third, not all the yields measured come from the rock. For example, the analyzed yields include tool background, as well as hydrogen and chlorine from both borehole and formation fluids. To get just the rock elemental concentrations, the undesired yields are left out of the oxides closure model, and the remaining yields are normalized to unity:

$$F\left(X_{Si}\frac{Y_{Si}}{S_{Si}} + X_{Ca}\frac{Y_{Ca}}{S_{Ca}} + X_S\frac{Y_S}{S_S} + X_{Ti}\frac{Y_{Ti}}{S_{Ti}} + X_{FeAl}\frac{Y_{Fe}}{S_{Fe}}\right) = 1,$$

(5-22)

where

F = normalization factor that compensates for the elimination of Cl, H, and tool background from the analysis and that the yields are relative and divided by their sensitivities

Y_i = relative yield of element i

S_i = sensitivity of element i to the prompt neutron capture measurement

X_i = oxide association factor used to convert the element to its appropriate oxide or oxide and related elements. For Si and Ti, this term simply converts Si to SiO_2 and Ti to TiO_2. For calcium, the model assumes that Ca primarily resides as $CaCO_3$, thus accounting for both the CaO and CO_2 terms in Eq. 5-20.

Using the oxides closure model, it is possible to compute the weight fraction, W, of each element in the formation from the relationship

$$W_i = F\frac{Y_i}{S_i}.$$

(5-23)

The resulting dry-weight elemental concentrations can be used to compute both mineralogy as well as matrix properties including matrix density, matrix sigma, and matrix neutron responses. For example, matrix density can be directly approximated as a linear combination of the elements silicon, calcium, iron, and sulfur with a standard error of only 0.015 g/cm^3 according to the relationship

$$\rho_{matrix} = 2.620 + 0.0490Si + 0.2274Ca + 1.993Fe + 1.193S,$$

(5-24)

where Si, Ca, Fe, and S are weight fractions of the elements silicon, calcium, iron, and sulfur as derived from the oxide closure processing. In pure quartz, substituting a value of 0.47 for the silicon weight fraction produces a matrix density of 2.65 g/cm^3.

The dry-weight elemental concentrations can also be used to calculate formation mineralogy through the solution of simultaneous equations or converted into the major mineral groups based on the SpectroLith* processing methodology (Fig. 5-59). Standard SpectroLith processing without the use of Mg delivers

clay = kaolinite + illite + smectite + chlorite + glauconite
carbonate = calcite + dolomite
anydrite = anhydrite + gypsum
pyrite = pyrite
siderite = siderite
coal = coal
QFM = quartz + feldspar + mica

based on the measured silicon, calcium, iron, and sulfur dry-weight elemental concentrations along with the derived aluminum concentration. The interpretation is based on an extensive, high-quality core database comprising chemical concentrations and Fourier transform infrared spectroscopy of the mineralogy measured on hundreds of sedimentary rocks.

The use of elemental capture spectroscopy and SpectroLith processing to determine the formation mineralogy is especially useful because it is conducted without reference to any other measurement and knowing the porosity is not required for the solution.

For well placement, individual elemental concentrations can be used for geochemical correlation between wells and as layer markers. Quantitative lithology volumes from capture spectroscopy significantly simplify porosity and subsequent saturation evaluation.

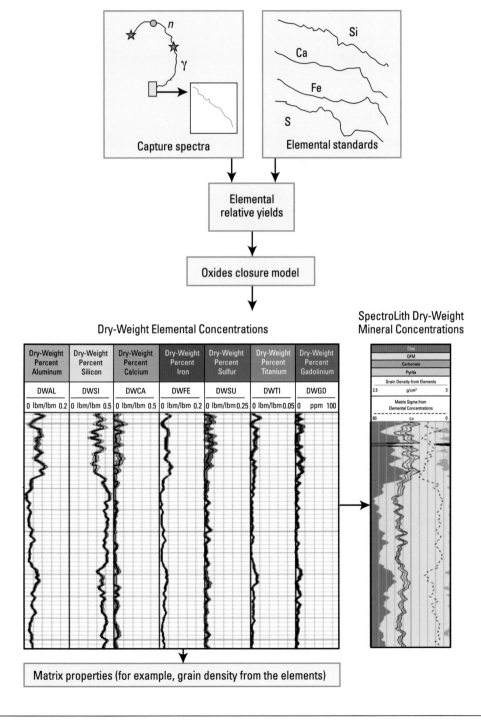

Capture spectra

Elemental standards

Si

Ca

Fe

S

Elemental
relative yields

Oxides closure model

Dry-Weight Elemental Concentrations

SpectroLith Dry-Weight
Mineral Concentrations

Dry-Weight Percent Aluminum	Dry-Weight Percent Silicon	Dry-Weight Percent Calcium	Dry-Weight Percent Iron	Dry-Weight Percent Sulfur	Dry-Weight Percent Titanium	Dry-Weight Percent Gadolinium
DWAL	DWSI	DWCA	DWFE	DWSU	DWTI	DWGD
0 lbm/lbm 0.2	0 lbm/lbm 0.5	0 lbm/lbm 0.5	0 lbm/lbm 0.2	0 lbm/lbm 0.25	0 lbm/lbm 0.05	0 ppm 100

Clay

QFM

Carbonate

Pyrite

Grain Density from Elements

2.5   g/cm³   3

Matrix Sigma from
Elemental Concentrations

60   cu   0

Matrix properties (for example, grain density from the elements)

Figure 5-59. Neutron capture spectroscopy processing involves spectral stripping to derive the elemental relative yields followed by oxides closure to determine the dry-weight concentrations of each of the elements present. Formation lithology and matrix properties such as grain density can be computed from the dry-weight concentrations.

5.12 Formation thermal capture cross section (sigma)

Sigma is the macroscopic thermal neutron capture cross section of the formation. The measurement makes use of the fact that after slowing down to thermal energy, neutrons linger in the formation and the borehole for several hundred microseconds, undergoing multiple collisions with nuclei in the surrounding material. Their capture by formation (and borehole) nuclei results in the emission of one or more gamma rays from the resulting highly excited nuclei.

As the neutron population near the tool declines as a result of capture and drift of the neutrons increasingly farther away from the tool (diffusion), the neutron and gamma ray flux observed in the tool detectors decreases. The primary cause of the decrease lies in the decline of the neutron population because of neutron capture. The measurement is therefore sensitive to the presence of thermal absorbers in the formation surrounding the tool. Chlorine has a much higher capture cross section than most other common elements in well logging. Sigma therefore becomes a very good indicator of the chlorine concentration around the tool and thus of the formation fluid salinity (Fig. 5-60).

Figure 5-60. Sigma, the macroscopic thermal neutron capture cross section of the formation, is closely related to the formation salinity.

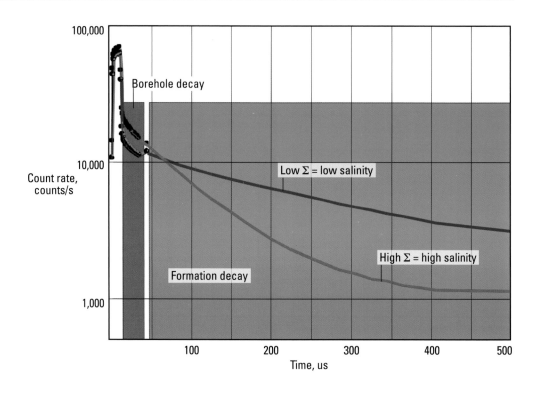

The sigma measurement is acquired by measuring the count rate of either neutrons or gamma rays in a single detector as a function of time. Initially the decrease in the count rate is due to the effect of the tool and borehole in proximity to the detector. The borehole effect is relatively small for sigma acquired using an LWD tool because the large LWD collar displaces most of the mud in the borehole. Sigma, Σ, expressed in capture units (cu), is related to the decay rate of the neutron or gamma ray counts in the detector:

$$\Sigma = \frac{4{,}550}{\tau},$$

(5-25)

where

τ = decay constant of the quasi-exponential decay of the neutrons or gamma rays (us).

Sigma is a volumetric measurement related to the chloride content of the formation. Because chloride generally occurs dissolved in the formation water, sigma can be used to derive resistivity-independent water saturation as long as the formation water is sufficiently saline to produce a usable sigma contrast between the water and hydrocarbons. This is a particularly useful measurement in formations where the traditional resistivity-based methods of estimating water saturation fail to provide reliable results (such as in some low-resistivity pay zones).

Typical sigma values for common formation components are in Fig. 5-61. There is a large difference between the sigma values for hydrocarbons and salty water.

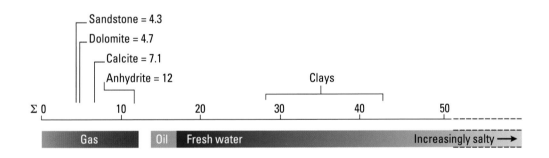

Figure 5-61. Typical sigma values for common formation components show a large difference between values for hydrocarbons and those for saline water.

Because sigma is a volumetric measurement, it responds to the relative volumes of the components in the formation according to the linear equation

$$\Sigma_{bulk} = \left(1-\phi\right)\Sigma_{grain} + \phi\,\Sigma_{fluid}.$$

(5-26)

Partitioning the fluid into water and hydrocarbon saturations yields

$$\Sigma_{bulk} = \left(1-\phi\right)\Sigma_{grain} + \phi\left[\Sigma_{water}\,S_w + \Sigma_{HC}\left(1-S_w\right)\right],$$

(5-27)

which can be transposed into a simple linear equation for water saturation:

$$S_w = \frac{\left(\Sigma_{bulk} - \Sigma_{grain}\right) + \phi\left(\Sigma_{grain} - \Sigma_{HC}\right)}{\phi\left(\Sigma_{water} - \Sigma_{HC}\right)},$$

(5-28)

where

S_w = formation water saturation

ϕ = formation porosity

Σ_{bulk} = measured bulk formation capture cross section

Σ_{water} = capture cross section of the water

Σ_{grain} = formation solid (grain) capture cross section

Σ_{HC} = hydrocarbon capture cross section.

In shaly formations sigma tends to correlate with the GR measurement as a result of the high capture cross section of clays. In clean carbonates it tends to anticorrelate with resistivity (Fig. 5-62, with the resistivity scale increasing to the right and the sigma scale increasing to the left). In a formation filled with salty water, such as that shown in the lower part of Fig. 5-62, sigma reads high because of the presence of chlorides in the water and the resistivity reads low owing to the conductivity of the salty water.

Where hydrocarbons displace salty water in the pores, the lack of chlorides in the hydrocarbons results in a decrease in the total chloride content of the formation and hence in the sigma measurement. The formation resistivity increases because the conductive water has been displaced by nonconductive hydrocarbons (middle interval in Fig. 5-62). The upper interval shows a low-resistivity pay zone composed of alternating thin beds of high and low resistivity. Because the beds are seen in parallel by the resistivity measurement, the measurement current prefers to follow the path of least resistance through the low-resistivity beds. If these thin bed effects are not taken in to account, S_w computed from the resistivity measurements indicates more water than is actually present in the formation. The sigma measurement is volumetric, which means that it is unperturbed by the path-of-least-resistance effect. That sigma reads low suggests the presence of more hydrocarbons in the formation than indicated by the resistivity measurements.

Sigma can be useful for well placement in low-resistivity pay zones (such as where resistivity anisotropy complicates conventional saturation determination) because saturation can be determined without use of resistivity measurements. Sigma can also be useful where shoulder bed and proximity effects complicate the resistivity response and hence conventional saturation calculations.

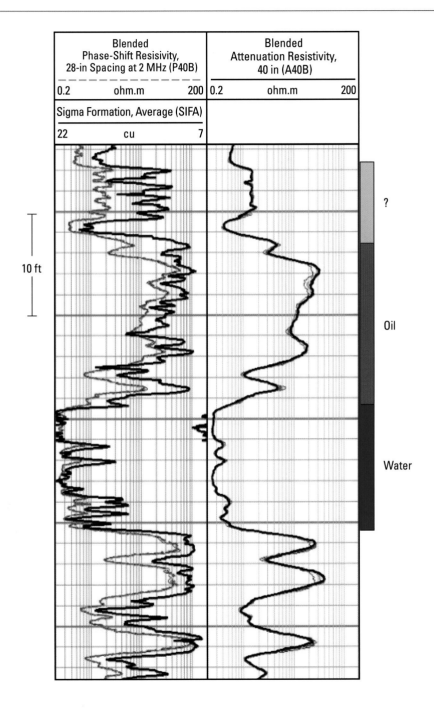

Blended Phase-Shift Resisivity, 28-in Spacing at 2 MHz (P40B)		Blended Attenuation Resistivity, 40 in (A40B)	
0.2	ohm.m 200	0.2	ohm.m 200
Sigma Formation, Average (SIFA)			
22	cu 7		

Figure 5-62. Sigma and resistivity anticorrelate in clean formations where hydrocarbons have displaced salty water. In the water interval, sigma is high and resistivity low. In the oil interval, sigma is low and resistivity high. The upper interval (orange) shows a low-resistivity pay zone in which the volumetric sigma measurement gives a better indication of the formation water saturation because it is not influenced by the path-of-least-resistance effect that complicates resistivity interpretation.

5.13 Formation pressure measurements

The fluid pressure in the pores of a formation can be measured by creating a hydraulic seal against the borehole wall, and then reducing the pressure on the formation and allowing the formation fluids to flow into a tool equipped with a pressure gauge. The movement of a pretest piston is generally used to drop the pressure in the tool flowline below formation pressure so that formation fluid flows into the tool (Fig. 5-63).

Figure 5-63. Acquiring a formation pressure measurement requires a seal between the tool and formation, a piston to reduce the pressure in the tool below formation pressure so that fluid flows into the tool, and a pressure gauge to measure the pressure response.

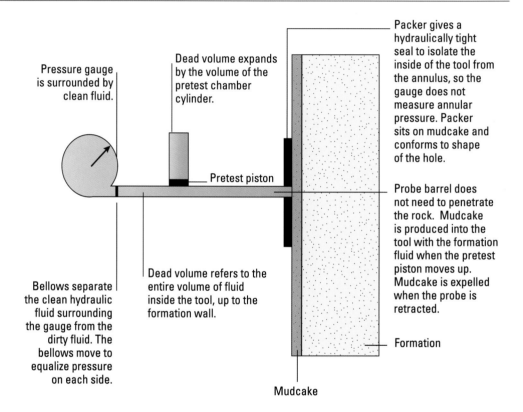

Pressure gauge is surrounded by clean fluid.

Dead volume expands by the volume of the pretest chamber cylinder.

Pretest piston

Bellows separate the clean hydraulic fluid surrounding the gauge from the dirty fluid. The bellows move to equalize pressure on each side.

Dead volume refers to the entire volume of fluid inside the tool, up to the formation wall.

Packer gives a hydraulically tight seal to isolate the inside of the tool from the annulus, so the gauge does not measure annular pressure. Packer sits on mudcake and conforms to shape of the hole.

Probe barrel does not need to penetrate the rock. Mudcake is produced into the tool with the formation fluid when the pretest piston moves up. Mudcake is expelled when the probe is retracted.

Formation

Mudcake

The sequence of formation pressure drawdown and buildup is called a pretest because it was originally designed to confirm a pressure seal between the tool and formation prior to making further tests and collecting samples. Today pretests are commonly used to acquire formation fluid pressure and mobility data only (Fig. 5-64). The drawdown and buildup rates provide information about how freely the formation fluid moves. The mobility of the fluid is defined as the ratio of the formation permeability divided by the fluid viscosity. This takes into account the ease with which the fluid moves (viscosity) and the ability of the rock to transmit fluids (permeability).

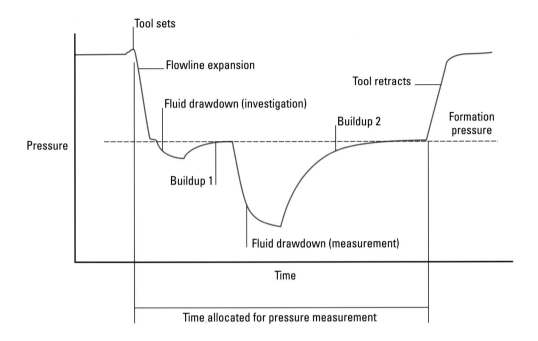

As shown in the example in Fig. 5-64, the pressure gauge initially reads the mud pressure in the borehole. There is a slight increase as the mudcake on the borehole wall is compressed by the sealing packer. The pressure in the flowline then drops rapidly as the pretest piston retracts. A small pressure anomaly occurs as the pressure in the flowline drops below formation pressure and the mudcake collapses into the flowline. At this point formation fluid begins to flow into the flowline and continues to do so when the pretest piston stops. The pressure builds until the pressure in the tool matches the formation pressure, at which point no further fluid flows into the tool. A second drawdown, for which the drawdown rate and volume are optimized based on the first drawdown and buildup, was conducted to improve the formation data acquired.

The hydrostatic pressure, p, of a fluid at depth is given by

$$p = \rho_{fluid} g_n (TVD),$$ (5-29)

where

ρ_{fluid} = fluid density

g_n = gravitational acceleration

TVD = true vertical depth.

One of the major applications of pressure data is for identification of the fluids and their contacts within the formation. In plots of pressure against the TVD of a well, fluids of similar density fall on a line (Fig. 5-65). The gradient of the line is related to the density of the fluid, which can be used to identify the type and density of the mobile fluid in the formation. Water, with a density about 1 g/cm^3, has the highest gradient, whereas oil and gas plot at lower gradients depending on their composition, temperature, and pressure. The intersection of the gradients can be used to infer the location of the fluid contacts in the formation, even if the well does not cross the contact. Formation pressure data between wells is used to identify reservoir compartments and uneven depletion.

Formation pressure data is important for reservoir management and well placement. Early identification of depleted zones enables pursuing alternative reservoir targets while the drilling equipment is still in the hole. Real-time knowledge of the formation pressure also facilitates mud weight optimization to ensure borehole stability while maintaining pressure control on the formation.

Figure 5-65. Plotting measured formation pressure against TVD can identify fluid gradients. The gradients indicate the density of the fluids (and therefore can help to identify the fluids), and the intersection of the gradients identifies fluid contact depths.

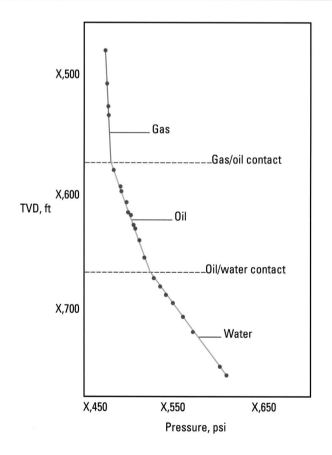

Formation pressure can be used to drill horizontal wells with greater accuracy. Given the uncertainties in surveying, wells that are intended to be horizontal may not actually be drilled horizontally. By using measurements of the formation pressure, adjustments to the well trajectory can be made to maintain a constant measured formation pressure, which indicates that the well is being drilled at a consistent TVD and is therefore horizontal.

5.14 Remote boundary detection

5.14.1 Seismic while drilling

Seismic-while-drilling services are configured with either the drill bit as a downhole energy source and receivers at surface (called drillbit seismic service) or a surface source with receivers downhole (Fig. 5-66). Both systems measure the one-way transit time of the seismic energy through the Earth. Because the depth of the downhole tool is known, the velocity of the seismic wave through the Earth can be determined. Knowledge of the seismic velocity through the various layers enables making a time to depth conversion of the surface seismic data to determine the depth of a feature identified on the surface seismic. In areas where knowledge of the seismic velocity through the layers is limited, real-time seismic data reduces the depth uncertainty of events ahead of the bit. The well trajectory can be altered by steering the bit to the appropriate depth to intersect geologic targets. The seismic data can also be used for pore pressure prediction ahead of the bit.

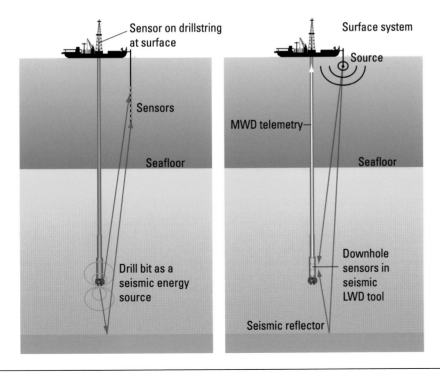

Figure 5-66. Seismic while drilling can be configured as drillbit seismic service (left), which uses vibrations from the drilling process as a seismic energy source. The drillbit seismic service can be used only under specific conditions. The more widely used seismic LWD system uses a conventional seismic source at surface and detects the signal downhole.

5.14.2 Electromagnetic distance to boundary detection

One of the major problems with traditional well placement is the limited depth of investigation of the measurements designed primarily for petrophysical formation evaluation. Although azimuthal data greatly enhances well placement by enabling determination of the direction from which features contact the borehole, most azimuthal measurements have a limited depth of investigation (of the order of inches). Effectively, this means that a boundary can be detected only when the borehole has come into contact with it when using conventional azimuthal measurements.

The development of deep directional electromagnetic measurements has revolutionized well placement by enabling the remote detection of resistivity changes within a formation. PeriScope bed boundary mapping service complements the LWD measurement portfolio by adding the azimuthal sensitivity previously only available with borehole images and extending the depth of investigation beyond that of conventional LWD propagation resistivity.

This deep directional capability bridges a gap in the real-time measurement portfolio (Fig. 5-67). Seismic data is able to visualize large structural features, but because of the limited frequency content of surface seismic data, some features are below seismic resolution. At the other end of the scale, conventional LWD images indicate the presence of a formation boundary, but because of their limited depth of investigation it is often too late to change trajectory in time to avoid exiting the reservoir. Conventional propagation resistivities have

Figure 5-67. PeriScope distance to boundary service delivers both directionality and depth of investigation at a useful scale.

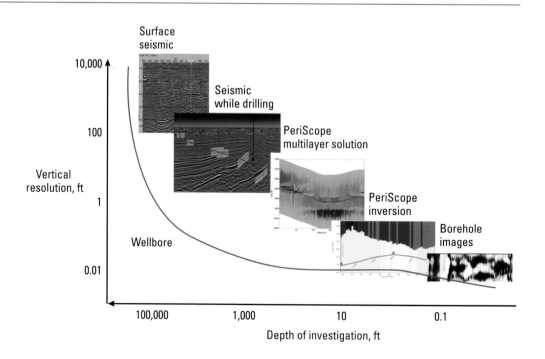

been used successfully for well placement relative to a boundary, but because these measurements are nonazimuthal, they do not indicate the direction from which a boundary is approaching the wellbore and hence do not facilitate informed well placement decisions. The PeriScope family of services, with its significant diameter of investigation and azimuthal sensitivity, delivers the directionality and depth of investigation required for advanced well placement.

5.14.2.1 Applications of remote boundary detection

The remote detection of surrounding layers opens up a whole new capability for well placement.

- Wells can now be positioned closer to the roof of a reservoir without risking exiting, ensuring that the minimum attic oil is left between the wellbore and reservoir roof when the reservoir is depleted.

- Wells can be drilled along meandering ancient river channels, detecting the banks, roof, and floor of the sand to keep the wellbore positioned in the productive channel.

- Wells can be drilled parallel to fluid contacts, enabling the exploitation of thin oil rims, for example.

- Events such as subseismic faulting can be identified and quantified. As thinner reservoirs are drilled, the evaluation of subseismic faulting becomes increasingly important because the borehole can exit the reservoir on encountering a fault.

- Evaluation of reservoir compartmentalization is improved through the ability to track the top and bottom of layers, thereby improving estimation of the volume of hydrocarbons in place.

- Determination of the resistivity in adjacent layers enables tracking water movement remotely. Experience has shown that a single well can track waterflood fronts in multiple layers simultaneously, reducing the number of observation laterals required to track water movement.

One of the key capabilities of remote boundary detection is providing early warning of changes in formation dip. Figure 5-68 shows a horizontal well encountering a change in formation dip. If a conventional propagation resistivity measurement with a depth of investigation of approximately 4 ft [1.2 m] is positioned 40 ft [12 m] behind the bit, the change in formation dip of 5.7° is not identified until the bit begins to exit the reservoir. Under the same conditions, the 15-ft depth of investigation of the PeriScope service gives sufficient early warning of a dip change of up to 20.6° before the bit exits the reservoir. The 15-ft depth of investigation of the PeriScope measurement gives 110-ft [34-m] MD advance warning to execute the trajectory change before the bit would exit the reservoir if the well continued to be drilled horizontally.

Figure 5-68. Remote detection of a change in formation dip can provide sufficient forewarning to enable trajectory adjustment and remain in the reservoir.

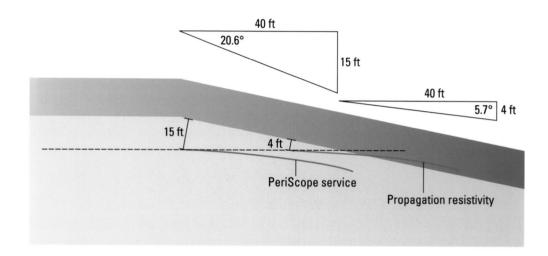

5.14.2.2 Measurement fundamentals

The PeriScope tool integrates two tilted receivers and one transverse transmitter with the conventional propagation resistivity array (Fig. 5-69). The conventional transmitters (red arrows) and receivers (blue arrows) can be used to deliver standard propagation resistivity measurements. The directional measurements used for the remote detection of formation resistivity changes are derived from the tilted receivers (black arrows). A transverse transmitter, T6 (green arrow), is used in conjunction with the axial and tilted receivers to derive resistivity anisotropy measurements at any well angle.

The increased transmitter-receiver spacings (34, 84, and 96 in [0.86, 2.13, and 2.44 m]) and operation at lower frequencies (100 kHz, 400 kHz, and 2 MHz) than conventional propagation resistivity measurements gain greater depth of investigation for the PeriScope measurement. Both directional phase-shift and attenuation measurements are made.

The tilted receiver coils create directional sensitivity at a 45° angle into the formation. The conventional axial coils do not show any sensitivity to tool orientation as the tool rotates. With tilted coils the receiver voltage depends on the orientation of the coil with respect to conductivity changes in the formation.

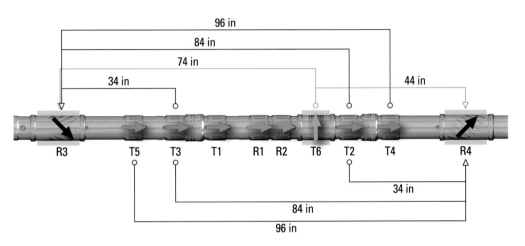

Figure 5-69. PeriScope deep directional measurements are acquired at three frequencies using axial propagation resistivity transmitters and directional receivers tilted relative to the axis of the tool.

The sensitive volume of a paired axial transmitter and tilted receiver is shown in Fig. 5-70. One side shows positive sensitivity and the other shows negative sensitivity.

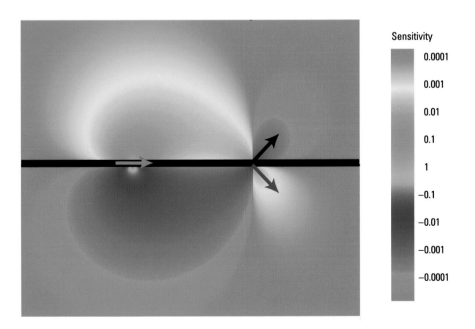

Figure 5-70. The sensitive volume of an axial transmitter and tilted receiver pair produces positive (upper) and negative (lower) lobes.

When rotated in a homogenous layer, the responses from the positive and negative lobes cancel each other, resulting in a net zero phase-shift and attenuation response. In a layered formation the directionality created by the tilted receiver results in a sinusoidal response. Figure 5-71 compares the phase shift at a tilted receiver to that at an axial receiver caused by excitation from an axial transmitter when the tool is placed close to a formation boundary. As the tool rotates, the axial receiver voltage is constant but the tilted receiver voltage shows sinusoidal variation with the tool rotational azimuth angle. The magnitude of the variation is related to the distance to the conductive boundary, and the signal maximum indicates the direction of a more conductive formation. This is the basic concept of the azimuthal measurements, providing both the direction and distance to the formation boundary.

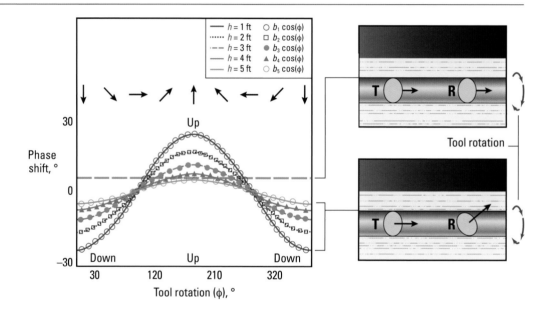

Figure 5-71. When placed near a change in formation resistivity, the signal in an axial receiver from an axial transmitter does not change with rotation of the tool. The signal in a tilted receiver varies sinusoidally with tool rotation. The magnitude of the sinusoid is related to the distance to the resistivity change and the resistivity contrast between the layers.

Symmetrization to eliminate dip and anisotropy effects

For homogeneous formations with boundaries parallel to the tool, interpretation of the directional responses is relatively simple. However, if the formation resistivity is anisotropic or there is a low incidence angle between the layering and tool, the interpretation becomes more complex. Figure 5-72 shows the response of a directional propagation tool in an anisotropic 20-ft [6-m] bed consisting of three layers: 3 ohm.m, R_h = 6 ohm.m and R_v = 30 ohm.m, and 1 ohm.m. The single axial transmitter and 45° tilted receiver pair (upper middle of Fig. 5-72) shows sensitivity to anisotropy including complex changes to the location of the maximum response. Symmetrization, much like borehole compensation, involves having the signal arrive at the receivers from above and below. The configuration of signals required to achieve this is shown in the upper right of Fig. 5-72. The resulting symmetrized pair has a significantly improved response independent of the resistivity anisotropy and dipping angle.

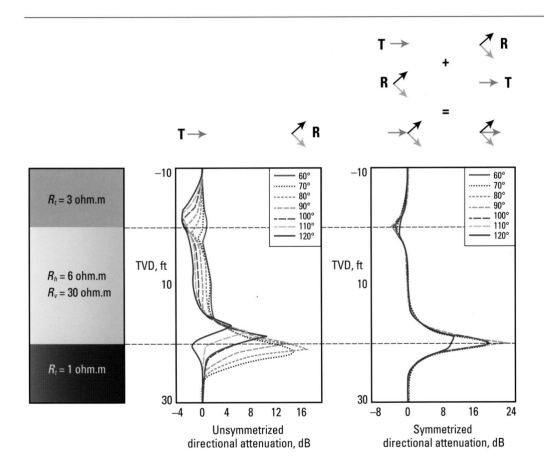

Figure 5-72. Symmetrization removes sensitivity to formation anisotropy and incidence angle, greatly simplifying interpretation of the directional responses.

An additional benefit of symmetrization is that the sensitivity map becomes symmetric between the transmitter and receiver, with the dominant contribution from the region between them (Fig. 5-73).

Figure 5-73. Symmetrization results in symmetric sensitivity between the transmitter and receiver.

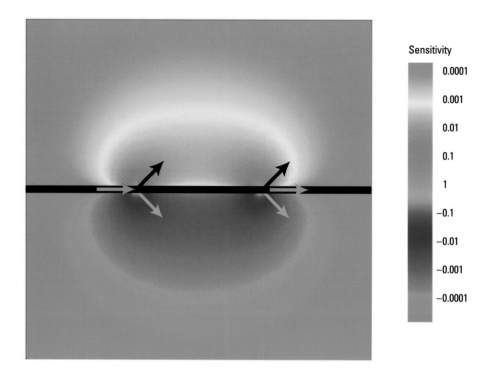

Tool response

When the tool is rotating, the symmetrized directional responses are plotted versus GTF (tool azimuth relative to the top of the hole). Figure 5-74 shows a cross section of the tool (solid circle on the right panel) rotating below a conductive bed, which is dipping at 60° in the cross section through the tool. The amplitude of the response varies sinusoidally, with the maximum amplitude occurring when the positive lobe is oriented toward the conductive bed. The minimum occurs when the positive lobe is oriented away (180° of rotational azimuth) from the conductive bed. On completion of an acquisition sequence, a sinusoid is fitted through the acquired points. This technique produces a more accurate determination of the azimuth of the bed than binning techniques, such as those used for density and resistivity images. Sinusoid fitting also creates a more robust measurement, able to tolerate irregular tool rotation and stick-slip during rough drilling operations.

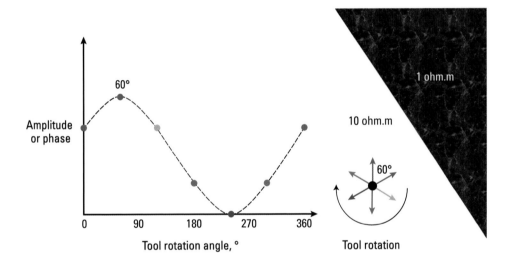

Figure 5-74. The symmetrized directional response is a maximum when the positive lobe is oriented toward the conductive bed and a minimum when oriented away. The azimuth of maximum amplitude found by sinusoid fitting enables determination of the azimuth of the bed.

The amplitude of the response depends on the distance to the resistivity change and the resistivity contrast:

- the closer the layer, the greater the amplitude
- the higher the contrast, the greater the amplitude.

If the layers have the same resistivity or the bed with contrasting resistivity is beyond the depth of investigation of the directional measurements, the amplitude of the sinusoid is zero.

The PeriScope directional response in the majority of cases is summarized in Table 5-7. The sign of the response (positive or negative) is controlled by the product of the location of the change above or below the tool and whether the bed is more or less resistive than the layer in which the tool is located.

- Beds more conductive than the layer in which the tool is located create a positive response.

- Beds more resistive than the local layer create a negative response.

- Resistivity changes above the tool create a positive response.

- Resistivity changes below the tool create a negative response.

Table 5-7. PeriScope Directional Response Summary

	Conductive +	Resistive −
Above +	+	−
Below −	−	+

As shown in Table 5-7, a positive directional response can be due to a more conductive bed above the tool or a more resistive bed below the tool. Similarly, a negative response can be due to a more conductive bed below the tool or a more resistive bed above the tool. This is resolved by using multiple directional measurements with differing depths of investigation.

The signal from a single bed can be thought of as the product of the conductivity contrast multiplied by the proximity of the bed to the tool. The total directional signal amplitude measured by the tool is the sum of the response from the positive and negative lobes. In situations where the signals from above and below the tool have the same amplitude, the total directional signal is zero because the positive and negative responses cancel.

These concepts can be explored by reviewing the tool response to the three-layer formation in Fig. 5-75. In the bottom panel, the three layers of different resistivity are traversed by a well (green line) inclined at 85° (5° incidence angle to the horizontal layering). The middle panel shows the conventional propagation phase (green) and attenuation (blue) resistivity responses. As each boundary is crossed, the propagation resistivities show positive polarization horns. These horns are useful for identifying the location of the boundaries, but they do not indicate whether the boundary is approaching from above or below.

The upper panel shows the PeriScope directional attenuation response. This response is not a resistivity but an attenuation measurement in decibels. The maximum amplitude of the response occurs where a boundary is crossed. Starting from the left, the attenuation signals are zero when the tool is far from the boundaries. When the 20-ohm.m layer is within the depth of investigation of the deeper attenuation measurement, the response becomes positive because the detected layer is below the tool (negative) and more resistive (negative) than the local layer. Negative multiplied by negative gives a positive response.

The response becomes increasingly positive until the tool passes through the boundary. Below the upper boundary, the response becomes less positive as the signals from above and below the tool partially cancel each other. When the tool is in the 20-ohm.m layer, the 2-ohm.m layer is relatively conductive (positive) and above the tool (positive) so the signal from the upper layer is positive. The lower layer is also relatively conductive (positive), but because it is below the tool (negative), the product of the positive and negative result in a negative signal from the lower layer. When in the 20-ohm.m layer, the relative proportion of the total signal resulting from the positive signal from above and the negative signal from below depends on how close the tool is to the respective layers. Just below the upper layer, the net response is positive because the positive signal from the upper layer dominates. Closer to the lower layer, the signal becomes negative as the lower layer dominates. At some point in the layer, the positive and negative signals balance, resulting in a zero net response.

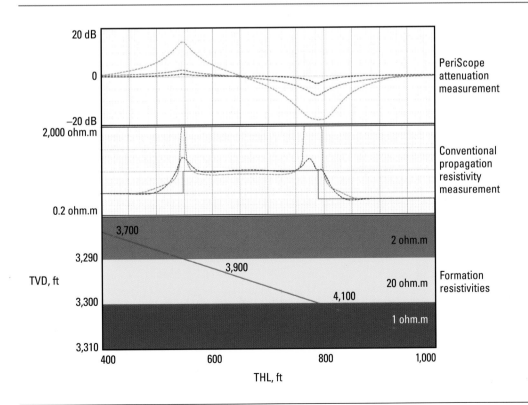

Figure 5-75. The response to a three-layer formation model (bottom panel) of conventional propagation resistivity (middle panel) shows positive polarization horns whereas the PeriScope directional attenuation measurement (top panel) develops a maximum response where a boundary is crossed, with the sign of the response controlled by the location of the layer above or below the tool and whether it is more or less resistive than the layer in which the tool is located.

The position of the zero response is slightly closer to the less conductive (the more resistive) of the two layers because the signal from this layer is weaker than the signal from the more conductive layer. For the two signals to be of equal amplitude and hence cancel, the tool must be closer to the less conductive layer. In a layer between two layers of equal conductivity, the zero point is in the geometric middle of the local layer.

As the tool passes into the 1-ohm.m layer, the signal amplitude remains negative because the 20-ohm.m layer is more resistive (negative) and above the tool (positive), resulting in a negative response. The response trails away to zero as the tool moves farther from the 20-ohm.m bed until it reads zero when surrounded by homogeneous 1-ohm.m formation within the depth of investigation of the tool.

Measurement naming

The conventional propagation resistivities from the axial transmitters and receivers are named in the same manner as normal arcVISION resistivities. For example, the attenuation resistivity acquired with 34-in transmitter-receiver spacing at 400 kHz is called A34L, and the phase-shift resistivity acquired with the same spacing but at 2 MHz is called P34H.

The names of the directional measurements used for distance to boundary determination have four components:

- symmetrized (S) or antisymmetrized (A)
- phase (P) or attenuation (A)
- shallow (S), medium (M), or deep (D)
- 100 kHz (1), 400 kHz (4), or 2 MHz (2).

Examples of the four-component measurement names are as follows:

- SAD4—symmetrized attenuation deep, 400 kHz
- SPS1—symmetrized phase shallow, 100 kHz
- APD2—antisymmetrized phase deep, 2 MHz.

The antisymmetrized data is used for anisotropy calculations and should not be used for distance to boundary inversions until after further development.

The boundary orientation with respect to the top of the hole is called

- DANG—boundary orientation (azimuth) derived from the 100-kHz and 400-kHz measurements
- HANG—boundary orientation (azimuth) derived from the 400-kHz and 2-MHz measurements.

Resistivity anisotropy measurements derived from the transverse transmitter (T6) start with the letters AN, followed by phase (P) or attenuation (A) and the frequency number (1, 4, or 2) as previously listed. These anisotropy measurements should not be used in distance to boundary inversions until after further development.

5.14.2.3 Three-layer model

The interpretation of PeriScope measurements is generally performed using a three-layer model (Fig. 5-76). In its most complex form the model has six unknowns:

- R_h—horizontal resistivity of the local layer

- R_v—vertical resistivity of the local layer

- R_u—resistivity of the bed above

- R_d—resistivity of the bed below

- h_u—distance to the bed above

- h_d—distance to the bed below.

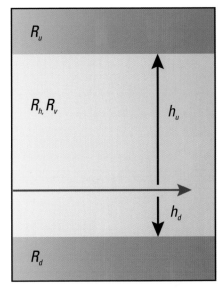

Figure 5-76. PeriScope measurement interpretation is based on a three-layer model with up to six unknowns. The red line represents the wellbore.

Because of symmetrization of the directional measurements, the wellbore incidence angle to the layers and resistivity anisotropy of the beds above and below do not influence the measurements and hence do not need to be known or determined as part of the interpretation.

The local layer resistivities, R_h and R_v, are determined by using conventional propagation resistivity measurements available from the axial transmitters and receivers on the tool. Solving for local-layer resistivity anisotropy requires both conventional phase and attenuation resistivity measurements. The resistivity anisotropy in the local layer is determined only to account for its influence on the conventional propagation measurements. The directional measurements are unaffected by anisotropy in either the local or remote beds.

It is recommended that one conventional attenuation and one conventional phase resistivity be used in inversion processing. Multiple conventional phase resistivity inputs are not recommended because the inversion model can solve only for anisotropy effect on the conventional measurements; it cannot account for invasion effects. If the conventional phase resistivities separate owing to the invasion effect, the inversion model tries to explain the separation by increasing the apparent formation anisotropy, resulting in an inaccurate assessment of the local layer resistivity. Inaccurate local layer resistivity reduces the accuracy of the distance to boundary evaluation.

The remaining four unknowns must be solved for by using directional measurements. A minimum of four directional measurements is required, though it is recommended that six be used if the real-time telemetry bandwidth is sufficient.

5.14.2.4 Crossplots

Crossplots are a way of visualizing the sensitivity of the PeriScope measurements to various conditions. They can be used to determine which of the array of measurements gives the best sensitivity to the expected formation resistivities. Appropriate crossplots should be generated before a PeriScope job so that they can be used in place of the automatic inversion if problems arise with Internet connectivity or the computer running the inversion processing.

A crossplot is a 2D representation of the data, with four of the model unknowns fixed, which leaves two to be presented on the crossplot. The most common crossplot configuration is to assume that the resistivities of the beds above and below the well are known and that the local layer thickness is also known. This reduces the problem to solving for the local isotropic resistivity and the distance to one of the boundaries. The distance to the other boundary is known because it is the bed thickness minus the distance to the opposite boundary.

In the formation model shown in Fig. 5-77, the beds above and below the 20-ft- [6-m-] thick reservoir layer have resistivities of 2 ohm.m and 1 ohm.m, respectively.

Figure 5-77. A formation model representing a 20-ft-thick isotropic reservoir layer between an upper bed of 2-ohm.m resistivity and a lower bed of 1-ohm.m resistivity.

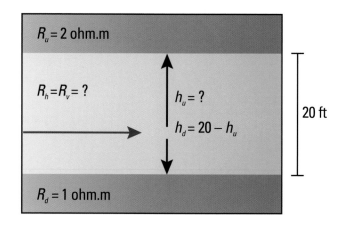

To keep the crossplot as simple as possible, a deep directional response is plotted against a conventional propagation resistivity. The conventional measurement helps define the resistivity of the local layer. The directional measurement is used to determine the distance to the boundaries.

Figure 5-78 shows one of many possible crossplots that can be made for the formation model in Fig. 5-77 by using two of the multiple measurements from the PeriScope tool. The conventional P28H resistivity is plotted on the x-axis because it is relatively shallow and hence responds primarily to the resistivity of the local layer. The SAD4 directional attenuation measurement is plotted on the y-axis because it is a deep measurement with good sensitivity to the adjacent beds.

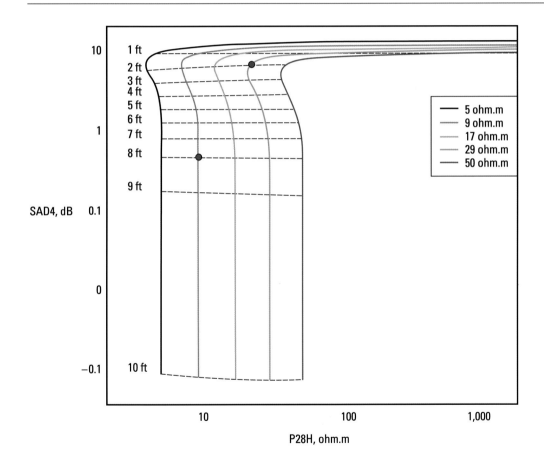

Figure 5-78. This PeriScope response crossplot is for the formation in Fig. 5-77.

The solid lines represent lines of R_t. The dashed lines represent the distance to the upper boundary. For example, if the measured P28H is 9 ohm.m and the SAD4 is 0.5 dB, plotting these values on the crossplot (red dot) reveals that R_t equals 9 ohm.m (purple solid line) and the upper boundary is 8 ft [2.4 m] above the borehole. Alternatively the plot can be used to deduce tool responses. For example, when 2 ft [0.6 m] from the upper boundary in a 29-ohm.m reservoir (blue dot), P28H is 21 ohm.m and SAD4 is 7 dB. These measurements indicate that 2 ft from the upper conductive layer, the phase 28-in [71-cm] transmitter-receiver resistivity measured at 2 MHz is lower than the local resistivity because part of the response is coming from the lower resistivity bed above. Getting any closer to the upper bed results in the P28H reading considerably higher because it polarizes close to the boundary.

The distance to the lower boundary is the 20-ft thickness of the reservoir layer minus the distance to the upper boundary. The 10-ft [3-m] distance to the upper boundary corresponds to a negative SAD4 response. This is because the lower bed is more conductive, which creates a stronger response in the directional receivers. To have the signals from the upper and lower beds balance, the tool must be slightly closer to the less conductive upper bed. In this case the directional zero point occurs at a distance of approximately 9.5 ft [2.9 m] to the upper bed.

The crossplot shows that the distance to the boundary is only weakly related to the local reservoir resistivity because the local resistivity has a range of only 5 to 50 ohm.m. If the local resistivity equaled the upper layer resistivity of 2 ohm.m, the distance would become a strong function of the local resistivity and ultimately the lines would collapse to a point because PeriScope service is unable to determine the distance to the boundary if there is no resistivity contrast with the target layer.

Crossplotting responses is a valuable technique for helping understand how the tool responds in a wide variety of conditions.

If data communication was lost to a rig running PeriScope service in the formation modeled in Fig. 5-78, drilling could continue with the Schlumberger field engineer adjusting the trajectory based on the SAD4 response. Depending on the desired distance to the upper boundary, the well could be steered up if the SAD4 response dropped below a target value and steered down if the SAD4 value became too large. For example, if the desired distance to the upper boundary was 9.5 ft, then the well could be steered to maintain an SAD4 value of zero. If the SAD4 response became positive the well placement recommendation to the client would be to drop inclination slightly. If the SAD4 became negative (indicating getting closer to the lower boundary) the well placement recommendation to the client would be to build trajectory inclination slightly.

5.14.2.5 Three-layer inversion fundamentals

PeriScope inversion automates the distance to boundary and remote resistivity determination with the added benefit that it takes into account all the data rather than only two channels from the array of measurements, as is the case with crossplotting data. The inversion uses a misfit-minimization technique in which it adjusts the three-layer model to find the set of six model parameters that best explains all the measured PeriScope responses. The output resistivity and distance to the layer above, resistivity and distance to the layer below, and resistivity of the layer in which the well is being drilled are displayed in real time as a color-coded resistivity cross section of the formation along the wellbore (right panel in Fig. 5-79).

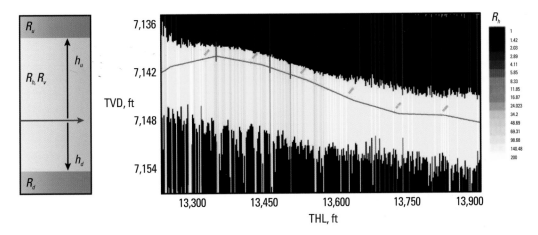

Figure 5-79. PeriScope inversion solves for the resistivity and distance to layers above and below the wellbore based on a three-layer model (left). The resulting information is displayed in real time as a color-coded resistivity cross section of the formation (right).

The inversion is calculated on a point-by-point basis. The results from one point are not used to guide the results for subsequent points. The independence of the points means that trends are most likely real and not a consequence of one point affecting the next.

As discussed previously, the PeriScope tool provides an azimuth to the boundaries around it, based on the assumption that the layers above and below are parallel. The boundary azimuth, in conjunction with distance to boundary information, is presented in an azimuthal view (Fig. 5-80). Generally the boundary positions derived from the latest 10 inversion points are displayed with the most recent data displayed in the brightest color. This allows evaluation of the distance to boundary trend along the trajectory.

Figure 5-80. The azimuthal viewer provides a representation of the subsurface as if looking down the borehole (center) with the upper (yellow) and lower (blue) boundaries shown at the distance and azimuth around the borehole determined from the PeriScope measurements.

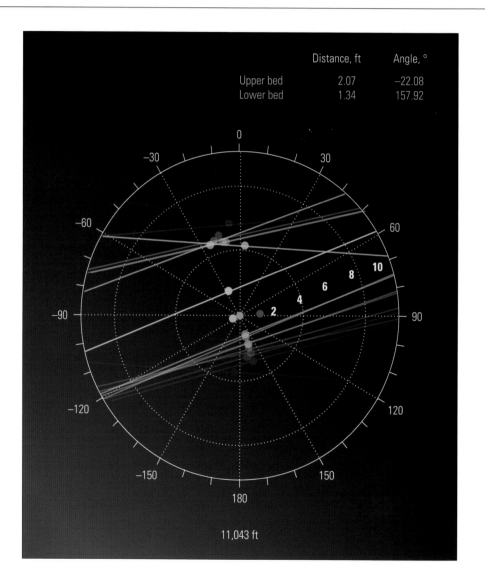

The distance and azimuth information enables steering a well in both TVD and azimuth relative to a resistivity boundary without having to come into contact with it. For example, in the situation shown in Fig. 5-79, the well could either be turned up to avoid the lower boundary or turned to the left, or a combination of the two depending on the most appropriate position for the wellbore in the target layer.

Returning to the three-layer example used previously, Fig. 5-81 adds the automatic PeriScope inversion result in this formation.

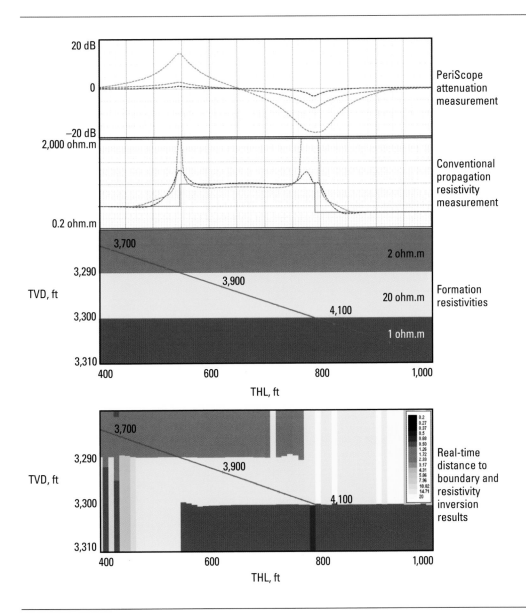

Figure 5-81. The automatic PeriScope inversion (bottom panel) solves for the three-layer model that best explains the measured PeriScope responses.

Starting from the left, the inversion finds the lower resistive boundary, but the response is not consistent because the directional signal from the tool is weak. The weak signal results from the conductive formation surrounding the tool, which limits the depth of investigation of the induced measurement currents in the formation. As the tool approaches the boundary, the inversion shows the boundary clearly because the signal becomes stronger. While the tool is in the 2-ohm.m layer the inversion cannot solve for the 1-ohm.m layer because the inversion is limited to solving for one boundary on either side of the tool. For this reason the 20-ohm.m layer appears to be infinitely thick when the tool is in the 2-ohm.m layer. Once the tool crosses into the 20-ohm.m reservoir layer, the upper and lower boundaries of the reservoir layer are visible. The lower boundary is not well defined at first because the signal from the 2-ohm.m layer dominates the PeriScope response. As the tool moves toward the center of the 20-ohm.m reservoir layer, the inversion is able to clearly distinguish the resistivities of the beds above and below the reservoir and the location of the two boundaries.

As the tool approaches the lower bed, the accuracy of the upper bed description deteriorates. The conductivity of the lower bed dominates the PeriScope response when the tool is close to it. Once the tool crosses the lower boundary of the reservoir, the upper boundary disappears because the inversion cannot solve for more than one boundary on each side of the tool. Although the location of the boundary between the lower bed and reservoir is reasonably well defined, the resistivity value for the reservoir is not consistent once the tool is in the lower bed. Again, the conductivity of the lower bed constrains the measurement currents around the tool. The higher resistivity of the reservoir layer does not create an attractive path for the measurement currents, which are seeking the path of least resistance. Thus the PeriScope measurements have relatively little sensitivity to the resistivity of the reservoir layer and consequently the inversion struggles to deliver a consistent value.

As can be seen from this example, the inversion response is closely linked to the physics of the PeriScope measurements. Where the measurements do not have the sensitivity to define a parameter, the inversion is not able to clearly define the reservoir. Understanding where the PeriScope measurements and inversion can and cannot provide answers is a vital part of proper prejob preparation.

Challenging environments

The basis of the PeriScope inversion on a three-layer model results in several challenging environments in which the inversion may deliver a cross section that is not a good representation of the subsurface.

One challenging environment for PeriScope inversion is where more than three layers affect the response of the tool. In thin layers, for example, the PeriScope measurements mix the responses from the various layers, with a bias toward the more conductive layers because the measurement currents seek the path of least resistance.

In the thin-layered formation model in Fig. 5-82, a horizontal well is drilled through a 100-ohm.m layer, over which are a 3-ft- [0.9-m-] thick 30-ohm.m layer and a 40-ohm.m layer. Below the 100-ohm.m layer is a 70-ohm.m layer (left side of the top panel). When water-flooded, the four layers are expected to have resistivities of 6, 3, 3, and 7 ohm.m from top to bottom (middle portion of the top panel). This effectively creates a three-layer formation. However, the 30-ohm.m layer may act as a boundary, preventing the lower layers from flooding, in which case the resistivity profile is as shown in the right side of the top panel. The PeriScope inversion results for these three resistivity profiles are shown in the bottom panel. The blue and red dots in the upper panel indicate where the PeriScope inversion identifies the boundaries between the layers.

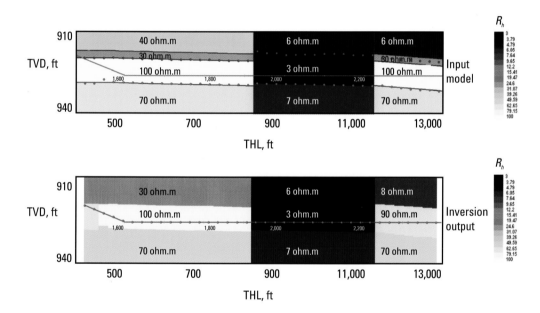

Figure 5-82. In thin layers, the PeriScope measurements respond to more than three layers, resulting in potentially misleading inversion results.

In the left and middle portions of the bottom panel, the PeriScope inversion does a good job of finding the correct boundary location and determining the resistivity on the other side of the boundary. However, the 40-ohm.m layer (left) is not identified because the more conductive 30-ohm.m layer dominates the signal. When the topmost layer is flooded, its resistivity drops to 6 ohm.m (middle and right). The PeriScope directional measurements respond to this conductivity and average it in with the 3-ft-thick 30-ohm.m layer. The location of the upper boundary is biased toward the conductive layer (blue dots upper right), and the resistivities of the layers are also biased. The inversion identifies the local layer as having a resistivity of 90 ohm.m rather than 100 ohm.m and the top layer as having a resistivity of 8 ohm.m rather than 6 ohm.m. In thin layers care should be taken to understand how the PeriScope tool and inversion combination responds to the layers and what boundaries can be identified.

Another challenging environment for PeriScope inversion is where resistivity changes in a formation do not show discrete steps. Transition zones between hydrocarbons and water are an example of where the resistivity can change smoothly from one value to another. The PeriScope responses in a ramp profile (Fig. 5-83) indicate higher resistivity above and lower resistivity below.

Figure 5-83. In a resistivity ramp profile, the PeriScope inversion represents the smooth resistivity change with a three-layer approximation because this is the model it solves for.

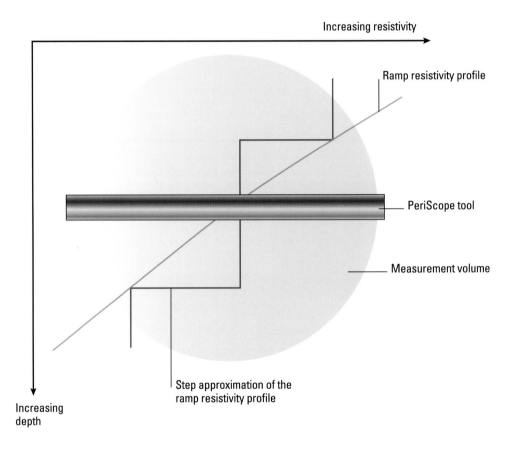

The inversion can solve for a three-layer model only, so it represents the smooth resistivity change as a three-layer sequence with higher resistivity above and lower resistivity below. If the borehole trajectory moves up or down in the ramp profile, the boundaries above and below the tool follow the trajectory. Inversion boundaries that appear to follow the well trajectory are a good indication that the well is being drilled in a resistivity ramp profile. When a sharp resistivity change is encountered either above or below the well, the corresponding inversion boundary no longer follows the well trajectory.

5.14.2.6 The future of distance to boundary technology

Distance to boundary technology is an area of rapid development. Since the introduction of the first electromagnetic remote detection-while-drilling technology in 2003, the capabilities of the service have advanced significantly. Recent areas of development include the following.

Multiple boundary detection

As outlined in Section 5.14.2.3, "Three-layer model," the basic inversion processing for PeriScope service assumes a three-layer model. This approach limits the number of boundaries that can be resolved to one above and one below the local layer of the well. By adopting a probabilistic approach to boundary detection, a multilayer solution can be derived from the PeriScope measurements.

The number of resistivity boundaries in the formation is not specified in the probabilistic approach. The multilayer inversion determines the probability of a boundary at each point above and below the well within the depth of investigation of the PeriScope measurements. The probability distribution is then plotted at subsequent points along the wellbore, creating a series of probability "wiggle" traces, as shown by the vertical black lines in Fig. 5-84. The probability wiggle traces can be interpreted in a manner similar to seismic wiggle traces. The location of a boundary is established where there is good coherence between high-probability peaks on the wiggle traces.

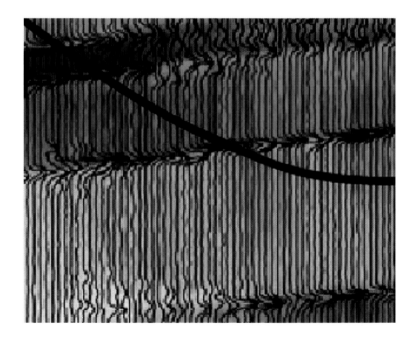

Figure 5-84. Probabilistic inversion techniques enable the detection of multiple boundaries above and below the wellbore (thick black line).

Color is used to identify the resistivity of the various layers, and the opacity of the color indicates the confidence in the resistivity evaluation. The closer to white the color becomes, the lower the confidence in the resistivity value at that point.

Starting from the left of Fig. 5-84 and following the well trajectory (thick black line) to the right, five layers are identified. When the tool is located in the uppermost green layer, the lower boundary of the layer is clearly defined while the positions of the three boundaries below the well become increasingly uncertain (the probability peaks are lower amplitude and wider) with distance from the well. The opacity of the color increases closer to the well, indicating increasing confidence in the inversion-derived layer resistivities near the well. As the well traverses through the layers, multiple coherent boundaries remain visible, with more sharply defined probability distributions closer to the well. The probability distribution of the upper boundaries widens and decreases in amplitude as the well moves farther away. This is due to the decreasing proportion of the directional measurement response from the increasingly distant layers.

Increased depth of investigation

The depth of investigation of electromagnetic measurements increases with lower measurement frequency and increasing transmitter-receiver spacing. By distributing directional transmitters operating at low frequencies along the BHA to gain increased spacing from the existing directional receivers of the PeriScope tool, very deep directional measurements (of the order of 100 ft [30 m]) can be made. Very deep measurements usually have multiple layers within their depth of investigation. Consequently the measurements must be interpreted with an inversion capable of distinguishing multiple layers above and below the borehole.

The application of the multiple-boundary inversion to the very deep directional measurements of the next generation of distance to boundary technology yields unprecedented real-time information about the geometry and resistivities of formation layering. Figure 5-85 shows an example application. During landing, the top of the target reservoir zone (upper yellow layer) is detected sufficiently early to allow adjustment of the trajectory to ensure that the well lands just below the top of the target layer. This guidance eliminates the need for a pilot well to identify the location of formation tops and ensures that the well landing in the zone is smooth, facilitating the installation of casing and completions. By knowing the location of the well with respect to the layering at all times, wellbore undulations are avoided. In turn, production is improved because water accumulation in troughs in an undulating trajectory and subsequent choking of hydrocarbon flow are avoided.

In addition to enabling steering the well at the top of the target layer, which minimizes attic oil above the well, the deep directional measurements can be used to identify the location of the oil/water contact (OWC). This facilitates well placement sufficiently far from the water and provides valuable information for reservoir volumetric calculations. On crossing a fault, the ability to identify multiple boundaries provides greater confidence in determining the location of the well. Consequently, well placement decisions to adjust the trajectory to place the well back at the top of the target layer can also be made with greater confidence.

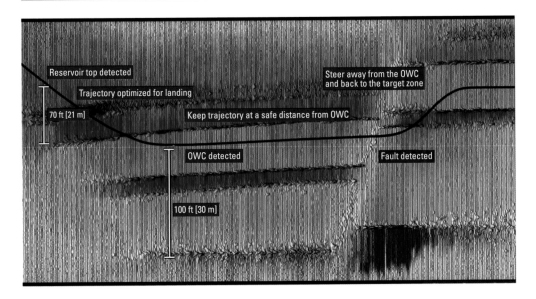

Figure 5-85. Increased depth of investigation, in conjunction with multiple-layer inversion capability, helps illuminate a wider path around the wellbore, revealing a larger perspective.

Information about the throw of the fault and the OWC in the new fault block helps further refine reservoir volumetric estimates. In this case, the different OWCs on either side of the fault suggest that the fault is sealed, preventing fluid migration across the reservoir. This is important information for developing accurate reservoir production models to optimize field production strategies.

Clearly a great amount of information can be extracted when multiple layers are illuminated above and below a well. The information is useful not only for real-time well placement optimization but also for static reservoir evaluation and dynamic reservoir production modeling.

Look-ahead capability

Look-around, look-ahead (LALA) capability has long been on the wish list of well placement professionals. The remote detection technologies outlined previously deliver look-around capability but do not provide information about formation changes ahead of the bit. Ongoing electromagnetic research suggests that look-ahead capabilities can be developed, enabling the identification of layering and fluid changes before the bit enters the different formation. For example, in low-inclination wells drilling can be stopped before an OWC is intersected. In high-angle and horizontal wells, look-ahead capability would enable the identification of faults, allowing TVD trajectory changes to be made in anticipation of fault throw or azimuthal trajectory changes to be made to avoid the fault if it is being approached obliquely.

Optimizing Well Placement for Productive Drilling

In an industry increasingly challenged to improve efficiency, productive drilling is essential. With rising well construction costs, more complex and marginal reservoirs, and limited asset technical expertise, delivering adequate return on investment becomes ever more difficult.

6.1 The two requirements for productive drilling: Penetration and placement

A high rate of formation penetration over extended intervals ensures that well construction costs are kept to a minimum while delivering maximum formation exposure. Placement is essential to ensure that the formation exposed is the target reservoir. A rapidly drilled well in the wrong place does not deliver the required production. An accurately placed well drilled excessively slowly incurs prohibitive construction costs. Fortunately, the use of appropriate well placement techniques can address both the speed and accuracy requirements.

Accurate well placement accelerates well penetration through

- reducing the need for pilot holes through remote identification of formation tops by deep-reading technologies

- reducing the need to redrill landing sections by ensuring smooth and accurate landing

- minimizing wellbore tortuosity, thereby easing casing and completion installation

- reducing the need for sidetracks to correct unanticipated exits from the reservoir

- improving average ROP by maintaining the wellbore within the reservoir layer, which is generally more porous and can be drilled faster than the surrounding formations.

Accurate well placement extends well penetration through

- minimizing tortuosity, thereby reducing frictional effects on the drillstring

- maintaining the wellbore within the reservoir layer.

Accelerated well construction enables

- earlier production and revenue generation

- more wells or greater reservoir exposure drilled per rig per year

- increased production.

6.2 The two consequences of productive drilling: Well performance and profitability

Accelerated well construction is only the tip of the iceberg of benefits. Often the most attractive advantage of accurately positioned wells is their higher production or injection performance. In the longer term, accurately placed wells are more likely to achieve better recovery of the hydrocarbons in place. As hydrocarbons become more difficult to find and produce, there is a growing need to improve recovery from existing accumulations to ensure their economic viability. Improved well performance, both in the short and long term, offers improved economics for reservoir development in new plays and brownfields alike.

Accurate well placement helps improve well construction efficiency, reduce drilling risk, extend reservoir contact, maximize reservoir exposure, improve well performance, enhance ultimate hydrocarbon recovery, and improve reservoir development economics.

6.3 Case studies

The following examples outline the techniques and benefits of accurate well placement.

6.3.1 Improving reservoir contact and production

Increasing borehole contact with the reservoir improves the production index (PI) of a well, resulting in more oil production per day for each unit of pressure drawdown between the borehole and the reservoir. Minimizing pressure drawdown helps avoid formation collapse, water coning, and gas cusping, the latter two of which result in production of unwanted fluids and potential bypass of oil.

A study[†] published by Saudi Aramco in 2003 shows the impact of well placement. Fourteen wells were reviewed, 10 of which had been drilled geometrically and 4 had been placed using real-time images. As summarized in Fig. 6-1, the improved proportion of the placed wells in the reservoir (net/gross ratio) and

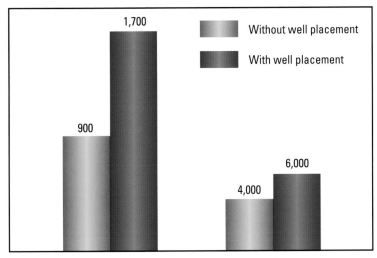

Figure 6-1. The introduction of well placement using real-time images to Saudi Aramco operations delivered a significant increase in the PI, substantially improving production for the same pressure drawdown compared with similar wells drilled geometrically.

Average Values	Without Well Placement	With Well Placement
Net/gross ratio, %	35	50
Total contact, ft	2,500	3,500
Net pay, ft	900	1,700
PI, bbl/psi-d	150	250
Rate, bbl/d	4,000	6,000
Number of wells	10	4

[†]Saleri, N.G., Salamy, S.P., and Al-Otaibi, S.S.: "The Expanding Role of the Drill Bit in Shaping the Subsurface," Distinguished Author Series, *Journal of Petroleum Technology* (December 2003) 55, No. 12, 53–56.

the ability to drill longer wells (total contact) resulted in a significant increase in the length of the reservoir exposed per well (net pay). The resulting improvement in the PI of the placed wells meant that the drawdown pressure difference could be reduced, minimizing the possibility of reservoir damage, while still increasing production by about 50%.

The payback period for the additional costs associated with the well placement was minimal considering the average production increase of 2,000 bbl/d. Improved return on investment is a common outcome in the comparison of wells steered using real-time well placement techniques with those drilled geometrically.

Development plans for a Middle East field called for drilling two 6-in-diameter horizontal laterals from the same 8½-in borehole (Fig. 6-2). The two laterals were to be placed in thin reservoir layers that had similar thicknesses and porosities—and structural uncertainties that made placement a challenge.

Figure 6-2. The upper lateral (blue), which was drilled using PeriScope imaging, was placed 100% in the target zone. That sharply contrasts with the lower lateral (green), which was drilled using conventional well placement techniques. Only 18% of that lateral was placed in the target zone, despite the availability of structural information acquired while drilling the upper lateral.

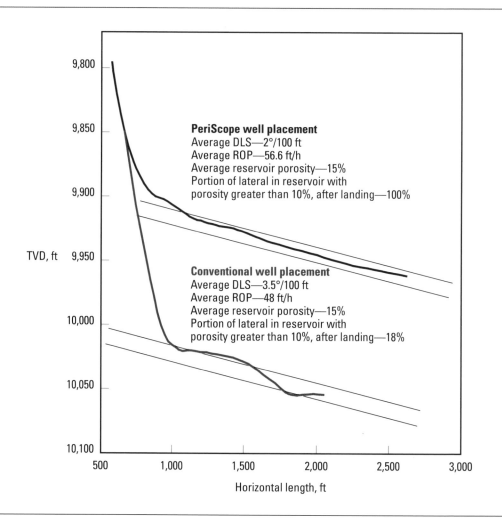

PeriScope well placement
Average DLS—2°/100 ft
Average ROP—56.6 ft/h
Average reservoir porosity—15%
Portion of lateral in reservoir with porosity greater than 10%, after landing—100%

Conventional well placement
Average DLS—3.5°/100 ft
Average ROP—48 ft/h
Average reservoir porosity—15%
Portion of lateral in reservoir with porosity greater than 10%, after landing—18%

The first lateral was drilled using PeriScope distance to boundary mapping technology to show the distance to the roof and floor of the layer. This enabled positioning the entire well within the highest-porosity portion of the reservoir by making gentle trajectory adjustments that minimized dogleg severity (DLS).

The second lateral was drilled using conventional LWD technologies and well placement techniques. This lateral was shorter than the one drilled using PeriScope imaging, had greater DLS, and provided less reservoir access, despite the advantage of having the carbonate formation's structure clarified by the first lateral.

The lateral drilled using PeriScope imaging was longer and less tortuous than the lateral drilled using conventional techniques, which enhanced production and improved casing and coiled tubing access (Fig. 6-3). That lateral was also drilled at a higher ROP than the conventionally drilled lateral, which reduced rig costs. The use of PeriScope distance to boundary imaging achieved more productive drilling and improved return on investment for the life of the well.

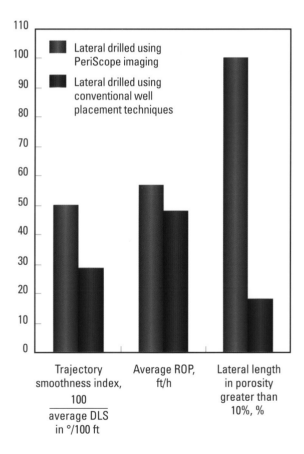

Figure 6-3. The use of PeriScope imaging dramatically improved well placement, compared with conventional techniques. It increased reservoir exposure for better production, cut costs by allowing the lateral to be drilled in less time, and reduced dogleg severity, which improved casing and coiled tubing access.

6.3.2 Maximizing reserves recovery

Drilling a well in the middle of a thick reservoir may leave behind significant volumes of unproduced hydrocarbon. Attic hydrocarbon may be trapped above the well as the water level rises, or cellar oil may be trapped below the well as a gas cap expands with decreasing reservoir pressure.

Well placement delivers long-term value by allowing wells to be positioned to maximize hydrocarbon recovery, in addition to addressing the shorter term issues of drilling ROP and reservoir contact.

A Russian operator wanted to maximize reservoir contact while keeping the well close to the cap shale. The complexity of the reservoir and the presence of randomly oriented calcite stringers or concretions made well placement using real-time image interpretation challenging. PeriScope distance to boundary mapping technology enabled remotely detecting the shale cap and steering the well as close as 8 in [0.2 m] from the reservoir roof without exiting to the shale. The 1,982-ft [604-m] horizontal section remained 100% in the reservoir, providing increased access to the oil in both the short and long term (Fig. 6-4).

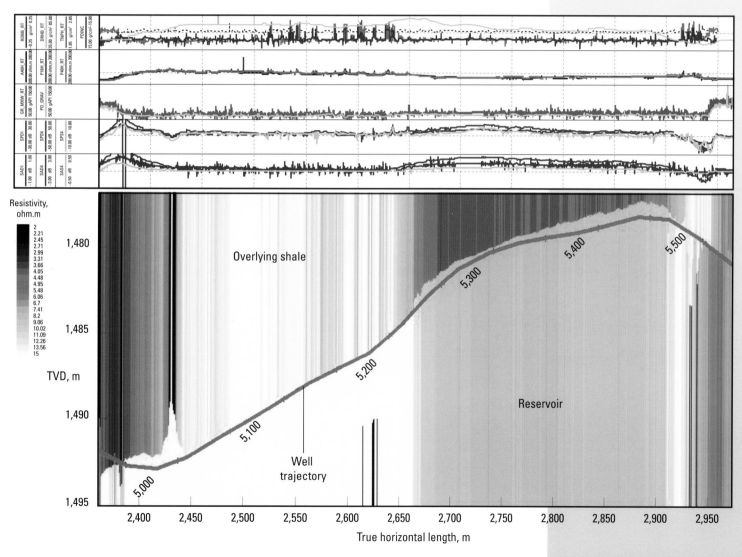

Figure 6-4. Use of the PeriScope distance to boundary mapping service enabled drilling the 604-m horizontal section close to the overlying shale and 100% within the reservoir, which increases access to the oil in both the short and long term.

6.3.3 Navigating thin layers

As operators target increasingly complex reservoirs, the ability to maximize contact with thin layers can have a substantial impact on asset economics.

An operator on Alaska's North Slope was targeting a sand layer estimated to be less than 10 ft [3 m] thick, sandwiched between shales. The area of interest was part of an anticline structure, with the well planned to be drilled approximately 6,600 ft [2,012 m] updip from the landing site to a major boundary fault. Because the sand was below seismic resolution, a simple geologic model was constructed in which the sand was assumed to be laterally continuous and unfaulted (Fig. 6-5).

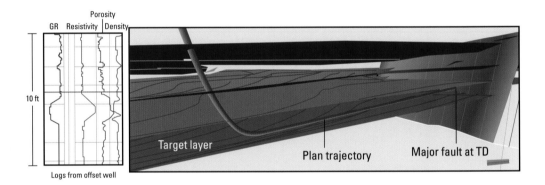

Figure 6-5. The planned well trajectory anticipated a uniform, 10-ft-thick, unfaulted target gently sloping up toward a major fault.

As drilling progressed it became apparent that the sand undulated, had several subseismic faults, and was as thin as 5 ft [1.5 m] in some sections. Navigating the thin sand required 13 trajectory adjustments, all guided by the PeriScope directional electromagnetic measurements. For example, the well was planned to drill up slope from the casing shoe based on an offset well correlation. However, within 30 ft [9 m] of the casing shoe it was observed that maintaining this inclination would result in the well exiting through the reservoir roof, so the well inclination was dropped to stay within the sand. Despite encountering two major faults and one minor fault, the well was successfully steered within the sand for 92.5% of the 6,678-ft [2,036-m] horizontal section (Fig. 6-6).

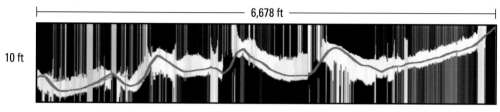

True horizontal length, ft

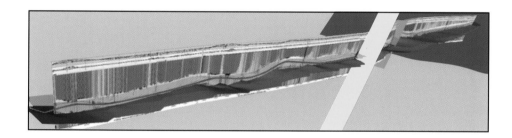

Figure 6-6. The actual well trajectory (red line, upper panel) had to avoid faults and bed undulations. The PeriScope deep directional electromagnetic measurements were inverted with the conventional propagation resistivities in real time to determine the distances to the shales above and below the well. Remedial action was taken to keep the trajectory in the sand or return to the sand when faults were encountered. A 3D view (bottom panel) shows the well successfully navigating 6,678 ft of the thin, undulating, faulted sand.

After evaluating the actual sand profile based on the PeriScope inversion results, the operator concluded that more than two sidetracks would probably not have been attempted if steering had used conventional techniques. This would have resulted in only 50% of the designed well length being drilled, most likely requiring an additional well to access the untapped reserves.

The ability to navigate thin reservoirs over long distances facilitates drilling of production and injection wells that deliver high recovery-factor displacement of hydrocarbons from small or isolated pools. This engenders confidence to develop targets previously considered uneconomical or involving excessive risk.

6.3.4 Azimuthal steering in meandering channel sands

Steering in thin sands is further complicated where the sands follow meandering channels in which they were deposited. In this situation identification of the roof and floor of the reservoir is not sufficient. Azimuthal information is required to facilitate steering of the well to the left and right to remain in the meandering channel.

A Middle East operator wanted to follow thin, sinuous channel sand stringers while minimizing wellbore tortuosity. Use of PeriScope distance to boundary results along with the deep azimuthal information identifying the direction of an approaching layer allowed the position of the well in the channel to be determined and the well steered, both in TVD and azimuthally, to remain in the sand (Fig. 6-7).

The well achieved 2,250 ft [686 m], corresponding to 82% of the horizontal section, in good-quality sand—an improvement of more than 50% over previous wells placed using conventional techniques. DLS was kept to less than 3°/100 ft, facilitating smooth completion.

Figure 6-7. Azimuthal steering using the PeriScope service provided a vertical cross section along the well length (lower left panel) and the distance and direction to approaching layers in an azimuthal viewer (upper left panel) as if looking along the wellbore in the direction of the bit. In this example the well is located near the lower right edge of the channel sand, as indicated by the close lower boundary (blue line) dipping to the left, which indicates that the thicker channel sand is to the left. A schematic of the well location in one of a sequence of meandering channel sands summarizes an interpretation of the data (right panel).

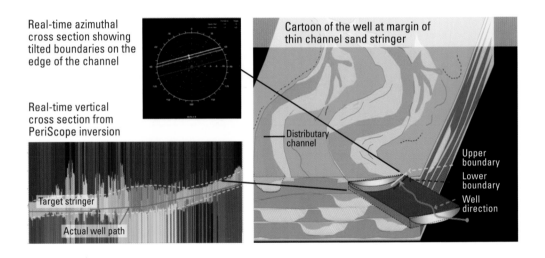

Real-time azimuthal cross section showing tilted boundaries on the edge of the channel

Real-time vertical cross section from PeriScope inversion

Target stringer

Actual well path

Cartoon of the well at margin of thin channel sand stringer

Distributary channel

Upper boundary

Lower boundary

Well direction

About the author

Roger Griffiths is a Schlumberger technical advisor in petrophysics and well placement. He is currently director of curriculum for LWD and well placement. He holds an honors degree in mechanical engineering from the University of Melbourne, Australia.